# SpringerBriefs in Environmental Science

DATE D

More information about this series at http://www.springer.com/series/8868

G. Thomas Farmer

# Modern Climate Change Science

An Overview of Today's Climate
Change Science

Springer

G. Thomas Farmer
Farmer Enterprises
Las Cruces, NM
USA

ISSN 2191-5547          ISSN 2191-5555   (electronic)
ISBN 978-3-319-09221-8          ISBN 978-3-319-09222-5   (eBook)
DOI 10.1007/978-3-319-09222-5

Library of Congress Control Number: 2014945257

Springer Cham Heidelberg New York Dordrecht London

Printed on acid-free paper

Springer is part of Springer Science+Business Media (www.springer.com)

An Overview and Update of Climate Change Science
(From January 2013 through February 2014)

Earthrise from the Moon, taken in 1968 by William Anders and Frank Borman as they circled the Moon. (From NASA, Public Domain)

*Perhaps the most important function of climate science on an issue of broad interest like global warming is to help educate the public and to provide useful input into the policy process. Governments, corporations, and individuals should listen to and learn from the science, just as intelligent people listen to their physicians when their health is in question. Good science input can inform wise policymaking. The role of scientists is to help assess the science and present it in an intelligible way that is policy relevant.*

The Forgiving Air: Understanding Environmental Change, (2008) by Richard Somerville, Distinguished Professor Emeritus at Scripps Institution of Oceanography, University of California, San Diego, CA, USA

# Preface

This book has been a challenge to write because I attempt what at first seems to be an impossible task because of two main causes for concern: (1) presenting the reader with a topical survey of the science of climate change and (2) directing the reader to the agencies and researchers at the forefront of the science; so why impossible?

(1) It is impossible to survey climate science adequately because it consists of an integration of complex subjects such as the physics and chemistry of the atmosphere, geophysics, geochemistry, geology, biology, ecology, biogeochemistry, computer science, paleoclimatology, and others. One small volume cannot do each of these topics justice in the context of "climate science" or make them a cohesive whole. However, there is an academic entity known as "climate change science" that is being studied today because the climate is changing. It is not the same as "climatology" or "climate science," it is the study of a changing climate, ergo climate change science.

(2) It is also impossible to direct the reader to all the agencies and researchers in the forefront of climate change science. There are simply too many of them.

The emphasis throughout the book is on research being conducted in the United States. This is because the writer is most familiar with the agencies and scientists doing work in the USA. Outstanding research institutions and individual researchers in other countries are included but the emphasis is on work being done at U.S. agencies and institutions.

There are two chapters in the book:

Chapter 1, an Overview of Climate Change Science, is a survey of what is changing with our current global climate. This chapter covers several topics, none of them in depth, but each with enough material that the reader is prepared to seek out and have an appreciation for a more advanced and thorough treatment of the subject.

Chapter 2, Status of Climate Change Research, leads the reader to sources of climate change information such as government agencies, colleges, and universities, and some of the leading researchers in climate change science.

The writer has kept the citing of URLs to a minimum because of the short lifespan of many of them.

Perhaps it is impossible to attempt to present such a wealth of information in one small volume. To attempt it has been a great challenge. Only the reader can judge as to whether it has been successful or not. This work may be considered a "bridge document" as it contains climate change information that updates Farmer and Cook's *Climate Change Science: A Modern Synthesis, Volume 1, The Physical Climate* published by Springer Publishers on January 12, 2013.

*Homo sapiens* is a destructive species. Since evolving around 200,000 years ago, it has set about on a course of destroying Planet Earth. It began using Earth's resources as soon as it learned it could tame fire to burn down areas of forest. As it learned to grow its own food it needed more land and the Agricultural Revolution was born around 12,000 years ago. It is not known when coal was first used as a source of energy but its use was vastly increased during the Industrial Revolution that began around 1750. *H. sapiens* discovered that coal, having taken millions of years to form, could easily be dug from its resting place in Earth, burned at Earth's surface, adding carbon dioxide to the active carbon cycle. Petroleum was made a popular fuel by the internal combustion engine and production went up for petroleum in the 1850s. Still more carbon was added to the carbon cycle from materials buried deep within the Earth. *H. sapiens* is continuing to disrupt the natural balance of the planet by burning fossil fuels, clearing forest lands, and manufacturing cement.

Las Cruces, USA, February 2014                                          G. Thomas Farmer

# Contents

# Figures

# Tables

# Acronyms and Abbreviations

| | |
|---|---|
| A&M | Agricultural and Mechanical |
| A2 | High growth emissions scenario |
| ACCMIP | Atmospheric Chemistry and Climate Model Intercomparison Project |
| AGU | American Geophysical Union |
| AIP | American Institute of Physics |
| AOGCMs | Atmosphere-Ocean General Circulation Models |
| AR | Assessment Report |
| AR5 | Assessment Report five |
| AVISO | Satellite altimetry data from CNES and NASA that measured ocean surface topography to an accuracy of 4.2 cm |
| BAMS | Bulletin of the American Meteorological Society |
| BER | U.S. Department of Energy Biological and Environmental Research |
| BEST | Berkeley Earth Surface Temperature |
| BGS | The British Geological Survey |
| BP | Before the Present |
| BPRC | Byrd Polar Research Center at Ohio State University in Columbus, Ohio |
| CAM | Climate Atmospheric Model |
| CAPS | Circumpolar Active-Layer Permafrost System |
| CARVE | NASA's Carbon in Arctic Reservoirs Vulnerability Experiment |
| CCS | Carbon Capture and Storage |
| CERES | Clouds and the Earth's Radiant Energy System |
| CICE | Climate Sea Ice model or program from the Geophysical Fluid Dynamics Laboratory, USA |
| CLIVAR | Climate Variability and Predictability Program |
| CLM | Climate Land Model |
| CMAR | CSIRO's Marine and Atmospheric Research |
| CMIP | Coupled Model Intercomparison Project |
| CNES | Centre National d'Etudes Spatiales, the space agency of France |

| | |
|---|---|
| COP | Conference of the Parties. Part of the UN Framework Convention on Climate Change |
| CPL | Coupler; software that combines two or more climate models or parts together |
| CRU | University of East Anglia's Climate Research Unit |
| CSIRO | Australia's Commonwealth Scientific and Industrial Research Organisation |
| DAMOCLES | Developing Arctic Modeling and Observing Capabilities for Long-term Environmental Studies |
| DC | District of Columbia (location of the capital of the U.S.) |
| DEC | December |
| DMS | Dimethyl sulphide |
| DNA | Deoxyribonucleic acid |
| DOE | U.S. Department of Energy |
| DOI | Digital Object Identifier is a character string ("digital identifier") used to uniquely identify an object such as an electronic document |
| EAIS | East Antarctic Ice Sheet |
| EBMs | Energy Balance Models |
| ECP | Extended Concentration Pathway |
| ECS | Equilibrium Climate Sensitivity |
| ENSO | El Niño-La Niña Southern Oscillation |
| EOR | Enhanced Oil Recovery |
| EPA | U.S. Environmental Protection Agency |
| ERB | Earth Radiation Budget |
| ESA | European Space Agency |
| ESMs | Earth System Models |
| ESRL | NOAA's Earth System Research Laboratory |
| EU | European Union |
| FAR | First Assessment Report |
| G-CISM | The Land Ice model or program, National Center for Atmospheric Research, USA |
| GCM | Global Circulation Model or Global Climate Model |
| GCOS | Global Climate Observing System |
| GCOS/GOOS | Global Climate Observing System/Global Ocean Observing System |
| GEUS | Geological Survey of Denmark and Greenland |
| GFDL | NOAA's Geophysical Fluid Dynamics Laboratory |
| GHCN | Global Historical Climate Network |
| GHCN-D | Global Historical Climatology Network-Daily |
| GHCN-M | Global Historical Climatology Network-Monthly |
| GHG | Greenhouse Gas |
| GISS | Goddard Institute of Space Studies |
| GISTEMP | Goddard Institute Surface Temperature |
| GLOBE | London-based Global Legislators Organisation |
| GODAE | Global Ocean Data Assimilation Experiment |

| | |
|---|---|
| GRACE | NASA's Gravity Recovery and Climate Experiment |
| GSC | Geological Survey of Canada |
| GSFC | NASA Goddard Space Flight Center |
| GSN | GCOS Surface Network |
| GtC | Gigatons of Carbon |
| $GtCO_2$ | Gigatons of Carbon Dioxide |
| GTNP | Global Terrestrial Network for Permafrost |
| GTS | Global Telecommunications System |
| HadCRUT | Hadley Centre Climate Research Unit—Temperature |
| HadSST | Hadley Center Sea Surface Temperature |
| HFC | Hydrofluocarbon |
| IARC | International Arctic Research Center |
| ICOADS | International Comprehensive Ocean Atmosphere Dataset |
| ICOS | Integrated Carbon Observation System |
| IHB | International Hydrographic Bureau, Monaco |
| IPCC | Intergovernmental Panel on Climate Change |
| IR | Infrared or longwave radiation |
| ITP | Institute of Tibetan Plateau Research in Beijing, China. Chinese Academy of Sciences' Institute of Tibetan Plateau Research (ITP) in Beijing |
| JPL | NASA's Jet Propulsion Laboratory |
| K | Kelvin temperature scale |
| LGM | Last Glacial Maximum |
| LLGHG | Long-lived greenhouse gas |
| LSCE | Laboratoire des Sciences du Climat et de l'Environnement |
| LST | Land Surface Temperature |
| MJ | Megajoule, $10^6$ joules |
| MODIS | Moderate Resolution Imaging Spectroradiometer |
| MOOC | Massive Open Online Course |
| MSU | Microwave Sounding Unit |
| NADW | North Atlantic Deep Water |
| NAO | North Atlantic Oscillation |
| NASA | National Aeronautic and Space Administration |
| NCAR | National Center for Atmosphere Research |
| NCDC | National Climate Data Center |
| NCIC | U.K. National Climate Information Centre |
| NIC | U.S. National Ice Center |
| NM | New Mexico |
| NOAA | National Oceanic and Atmospheric Administration |
| NSF | National Science Foundation |
| NSIDC | National Snow and Ice Data Center |
| NY | New York |
| NYU | New York University |
| OCB | Ocean Carbon and Biogeochemistry |
| OSU | Ohio State University in Columbus, Ohio |

| | |
|---|---|
| PCMDI | Program for Climate Model Diagnosis and Intercomparison |
| PD | Peak and Decline |
| PDO | Pacific Decadal Oscillation |
| PETM | Paleocene-Eocene Thermal Maximum |
| PFC | Perfluorinated Compounds |
| PFTBA | Perfluorotributylamine |
| POP | Part of the ocean program for climate modeling; National Center for Atmospheric Research, USA |
| QBO | Quasi-Biennial Oscillation |
| R-AL | Republican-Alabama |
| RCP | Representative Concentration Pathway |
| RF | Radiative Forcing |
| RGCM | Regional and Global Climate Modeling |
| RIGC | Japan's Research Institute for Global Change |
| RSS | Remote Sensing Systems |
| SAR | Second Assessment Report |
| SORCE | Solar Radiation and Climate Experiment |
| SRES | Special Report on Emissions Scenarios |
| SSAI/NASA | Science Systems and Applications, Inc. (SSAI). A major NASA contractor |
| SST | Sea Surface Temperature |
| TAR | Third Assessment Report |
| TERRA | NASA scientific research satellite in a Sun-synchronous orbit around the Earth |
| TERRA/MODIS | The satellite carrying MODIS |
| TIM | Total Irradiance Monitor |
| TIROS-N | Television Infrared Observation Satellite—North |
| TOA | Top of the Atmosphere |
| TOPEX | TOPEX/Poseidon is a joint venture between CNES and NASA |
| TOPEX/Poseidon | A joint venture between CNES and NASA |
| TPE | Pole Environment, an international program led by the |
| TSI | Total Solar Irradiance |
| U.C. | University of California |
| U.K. | United Kingdom |
| U.S. | United States |
| UAH | University of Alabama—Birmingham |
| UN | United Nations |
| UNEP | United Nations Environmental Programme |
| UNFCCC | United Nations Framework Convention on Climate Change |
| URL | Uniform Resource Locator (or web address) |
| USA | United States of America |
| USGS | United States Geological Survey |
| UV | Ultraviolet or shortwave radiation |
| UW | University of Washington |
| UWCRP | Under the World Climate Research Programme |

| | |
|---|---|
| VH | Very High |
| VL | Very Low |
| WAIS | West Antarctic Ice Sheet |
| WCRP's | World Climate Research Program's Intercomparison Project |
| WG | Working Group |
| WGCM | Working Group on Coupled Modeling |
| WMO | World Meteorology Organization |
| WOCE | World Ocean Circulation Experiment |

# Chapter 1
# Overview of Climate Change Science

**Abstract** A topical introduction to climate change, with emphasis on January 2013 through February 2014, is outlined in this chapter with a brief introduction to each topic. The reports of the IPCC AR4 and AR5 are briefly described and the IPCC and its history are discussed. The difference between weather and climate is explained with examples of each. Components of the climate system are illustrated. RCPs and ECPs are introduced, explained, and given in tabular form. The uncertainties as used by IPCC AR5 2013 are included. Earth's energy imbalance is discussed, illustrated, and explained as is its relationship to and cause of global warming. The range of possible climate sensitivities is given along with a discussion. The range of global temperatures is illustrated and discussed. Climate forcing and feedbacks are described and the differences between them are given with examples of each. Carbon capture and sequestration (CCS) is illustrated. The Keeling curve is shown and explained and carbon dioxide is discussed. An introduction to climate modeling is given and illustrated. Earth's melting ice is shown with examples and illustrations from permafrost, Antarctica, and Greenland.

**Keywords** Antarctica · Climate · Greenland · Representative concentration pathways · Energy · Imbalance · Sensitivity · Temperature · Components of the climate system · Paris · London · Iceland · Atmosphere · Carbon dioxide · Geosphere · Biosphere · Hydrosphere · Cryosphere · Global warming · RCPs · Modeling · Planet earth · Denial · Textbook · Moscow · Pakistan · Amazon · Australia · Weather · El Niño · La Niña · UN · United Nations · IPCC · FAR · SAR · TAR · AR4 · AR5 · Probability · WMO · World Meteorological Organization · Troposphere · Stratosphere · Scenarios · Celsius · Charles David keeling · Tipping points · Sea ice · keptical science.com · U.S. congress · Marcott · Keeling curve · Skin layer · Hansen · Glaciers · Ice caps · Climate change · NOAA · SRES · Atmospheric administration · UNEP · Obama · Herculean · 30-year time period · Decadal · W/m$^2$ · $CO_2$ · "Goldilocks planet" · GHGs · Greenhouse effect · Enhanced greenhouse effect · UV · IR · GtC · $GtCO_2$ · 3.67 · ECS · Roswell · NM · Radiative forcing · Mauna Loa · ENSO · PDO · World Ocean · Argo floats · Kuroshio Current · Permafrost

© The Author(s) 2015
G.T. Farmer, *Modern Climate Change Science*,
SpringerBriefs in Environmental Science, DOI 10.1007/978-3-319-09222-5_1

## 1.1  Introduction

There is no intention of covering each and every agency, individual, paper, idea, and book published during the past year (January 2013–February 2014) in climate change science. That would be a Herculean task and not within the scope of this work. It is, however, intended that this book emphasizes the most important results of recent research in climate change science. Each chapter ends with a list of references and further readings. Many of the leaders in their fields may be gleaned from these listings if they are not noted in the chapter. There is no doubt that some research that climate scientists deem important will be missed or purposefully left out and the writer can only apologize to those whose work is important but not included.

There are many scientists conducting significant work in the field of climate science today and their numbers are increasing at a rapid rate. This increase is as it should be, for the most important aspect of Earth and its inhabitants to be faced in the 21st century is climate change. Climate can change in many ways, but it is usually thought of as either warming or cooling. The planet is warming and this fact is unequivocal. How much it will warm is unknown; it depends on human action or inaction but we have a good idea of the range in temperature we can expect during this century, between 1.5 °C on the low end and perhaps 6 °C on the high end. The future conditions for living organisms (including humans) existing on Planet Earth will likely be determined within the next few decades by the work of scientists, engineers, and policy makers and the choices they make.

Denial of the conclusions of climate science and the disturbing role of oil and coal companies' concerted efforts to deny the science and delude the public, and the movement away from educating students in scientific subjects and mathematics can spell disaster for humankind. A 4.5–6 °C increase from where we are now or from the Industrial Revolution (around 1750) will make Earth uninhabitable for humans. If we reach plus 4.5° from the present global temperature, an additional few degrees won't make any difference; the planet will no longer be habitable for humans.

Many of the world's humans are observing the effects of climate change now. Those who are experiencing climate change first hand, such as some coastal communities and Pacific islanders, have no doubt about the fact that the globe is warming and sea level is rising.

Though natural climate cycles (i.e., El Niño, La Niña) may have slowed the atmosphere's rising temperature since 1998, 2012 was one of the 10 hottest years since 1880, according to the report released on August 6, 2013 by the U.S. National Oceanographic and Atmospheric Administration (NOAA).

The first decade of the 21st century was warmer than the 1990s, the 1990s were warmer than the 1980s, and the 1980s were warmer than the 1970s. Earth is warming.

The main components of Earth's climate system are seen in Fig. 1.1 and are discussed later in the text.

In the 1820s Jean Baptiste Joseph Fourier calculated that an object the size and shape of the Earth and its distance from the Sun should be considerably colder than

**Fig. 1.1** Components of Earth's climate system (from John Mason, www.skepticalscience.com, October 2013)

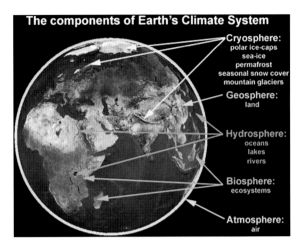

the planet actually is. He examined various possible sources of the observed heat in articles he published in 1824 and 1827. Fourier's consideration of the possibility that the Earth's atmosphere might act as an insulator of some kind is widely recognized as the first proposal of what is now known as the 'greenhouse effect.'

Svante Arrhenius, a Swedish chemist, in 1896 was the first scientist to attempt to calculate how changes in the levels of carbon dioxide in the atmosphere could alter the surface temperature through the greenhouse effect. Arrhenius was the first person to predict that emissions of carbon dioxide were large enough to cause global warming.

Additional information on the history of climate science may be found at the American Institute of Physics (AIP) web site below. Dr. Spencer Weart, author of *The Discovery of Global Warming*, discusses the history of climate change research at the following website: http://www.aip.org/history/climate/author.htm.

Dr. Weart discusses the history of global warming science in essay form in the material at the above website, and it is enjoyable reading and highly recommended.

Earth's climate has changed before but it is not "always changing." The climate usually changes over hundreds and thousands of years. What makes the current climate change unique is the rapidity of the change and the "fingerprints" of humankind.

Humans have survived ice ages and at least one interglacial but the climate has never changed as rapidly as it is changing now. Figure 1.2 shows climate change relative to the evolution of *Homo sapiens* and the development of civilization.

Previous major global climate changes, those beginning about 3 million years ago, were cycles that began before humans evolved. The human species appeared only about 200,000 years ago. Our distant ancestors survived multiple gradual cycles of 'ice ages' and interglacials but they remained in tropical Africa and probably were not bothered by extremely cold temperatures.

There is evidence that the Earth had begun to cool before the current global warming began. It is likely that, without the anthropogenic greenhouse gases added to the atmosphere, the climate system could have been headed to a new ice age.

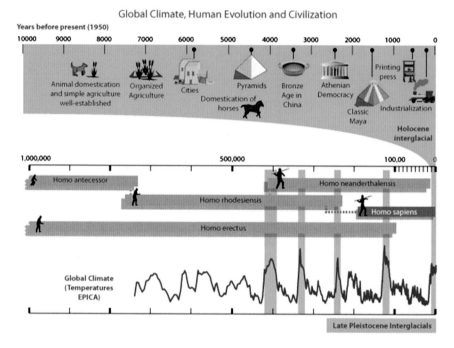

**Fig. 1.2** Global climate, human evolution, and civilization. Note the two time scales; the *upper scale begins on the left* at 10,000 years ago. The time scale in the *middle of the figure* begins on the *left* at 1,000,000 years ago (image by John Garrett; from www.skepticalscience.com)

Human civilization is roughly 12,000 years old as defined by the start of permanent settlements and agriculture. Agriculture became established as glaciers retreated from the last ice age. Modern society has developed entirely in our current geological Epoch, the Holocene (or the Anthropocene). Global temperatures haven't varied by more than ±1 °C since modern society developed. There have been regional shifts in climate (Medieval Warm Period, Little Ice Age, etc.), but since civilization began humans have never experienced a hotter global climate than today's climate.

Going further back in time, over a million years or so, our pre-human predecessors might have experienced a series of long cold glacial cycles. Several short interglacial periods were as warm as or slightly warmer than our current climate. For example, the climate 400,000 years ago was slightly warmer than now. But more typically for the last million years it's been 4–8 °C colder. Each transition from warm to glacial ages and back took thousands of years, giving humans and pre-humans many generations to adjust. So the climate hasn't changed much since humankind formed towns, villages, and cities, and irrigated farmland. Then in the mid-20th and into the early 21st centuries, Earth's global temperature began to increase rapidly.

In the next 100 years into the future our children and grandchildren and all future generations could be the first ever to experience a climate of as much as 6 °C or more above pre-industrial temperature but we seem to be headed there now. It is doubtful that Earth would remain habitable at such high temperatures.

## 1.2 Weather and Climate

Weather is the condition of the atmosphere from place to place over Earth's surface at present and within a week or so into the future (e.g., weather forecasts). Climate is defined as atmospheric conditions in a given area over at least a 30-year time period. Weather is short-term, climate is long-term. Climate is what one expects, weather is what one gets.

Weather and climate are often confused by those who don't understand the difference, including many meteorologists. For example, the statement commonly made by climate change/global warming deniers is the half-truth, "The climate is always changing." This statement is incorrect. Weather is always changing. Has the climate changed in Denver, Greenland, Paris, Miami, Iceland, Hong Kong, Manila, or London in memory? The climate in those places has not changed, except each has gotten a bit warmer. Climate is sometimes defined using the scientific definition: Climate is the statistics of weather over a 30-year period.

There are factors that affect climate that occur over the short term, such as El Niño and La Niña occurrences, positive and negative oscillations in ocean currents and others but most climate changing events occur over longer periods of time. For example, the buildup of greenhouse gases in the atmosphere. The buildup of carbon dioxide has been taking place over at least the past 260 years and it continues virtually unabated today.

Climate and factors that affect climate vary over decadal (decades), centennial (hundreds), millennial (thousands), and millions of years. It takes millions of years for geologic climate changes to occur, such as the origin of a mountain range, or hundreds of thousands to millions of years for a tectonic plate carrying carbonate rock to cycle through the crust. It also takes hundreds of thousands to millions of years to make coal and large quantities of oil.

## 1.3 The Intergovernmental Panel on Climate Change

The Intergovernmental Panel on Climate Change (IPCC) is unique. It is the foremost authority on climate change in existence. It consists of hundreds of volunteer climate experts from all over the world.

The IPCC is under the auspices of the United Nations (UN). It reviews and assesses the most recent scientific, technical and socio-economic information produced worldwide relevant to the understanding of climate change. It does not conduct any research nor does it monitor climate-related data or parameters.

The IPCC has produced reports on the status of Earth's climate beginning with the IPCC First Assessment Report 1990 (FAR). This was followed by the Second Assessment Report, Climate Change 1995 (SAR); the Third Assessment Report, Climate Change 2001 (TAR); and the Fourth Assessment Report, Climate Change 2007 (AR4). The Fifth Assessment Report (AR5) is being released over the September 2013 through October 2014 timeframe. The first AR5 volume, "*The*

*Physical Science Basis"* was published as a final draft and made available to the public on September 30, 2013. Reference to material from the 2013 AR5 report is included in this work.

The IPCC is an extraordinary undertaking. Hundreds of scientists volunteer to synthesize the findings of thousands of peer-reviewed scientific articles to provide policymakers and the public with the best current scientific understanding of climate change. Hundreds more experts review and provide comments on sections of the report. The IPCC is intergovernmental, which means that governments that are members of the UN can input their ideas to the IPCC reports. However, the reports are written by scientists and the final language is that used by scientists, not politicians or policy makers.

The Fifth Assessment Report (AR5 2013) documenting the 'physical science basis' of climate change is a magnificent scientific document. This report does not deal with fires, floods or hurricanes; those topics will be covered in a companion report on "impacts, adaptation and vulnerability" to be released in March 2014 and written by Working Group II. A report to be released in April on "mitigation of climate change" written by Working Group III will include renewable energy, energy efficiency and reduced deforestation; a synthesis report will be released next October 2014.

The IPCC 2013 report is over 2,200 pages. It's an impressive document but not the most accessible or the most readable. There are a few main points from the IPCC report released 30 September 2013 as given below. The quoted text is taken directly from the report's 36-page Summary for Policymakers.

1. Climate change is occurring now and will continue for the foreseeable future. "Warming of the climate system is unequivocal, and since the 1950s, many of the observed changes are unprecedented over decades to millennia. The atmosphere and ocean have warmed, the amounts of snow and ice have diminished, sea level has risen, and the concentrations of greenhouse gases have increased."

2. The climate changes we are witnessing now are unprecedented in human history. "The atmospheric concentrations of carbon dioxide ($CO_2$), methane, and nitrous oxide have increased to levels unprecedented in at least the last 800,000 years (based on ice core records). $CO_2$ concentrations have increased by 40 % since pre-industrial times, primarily from fossil fuel emissions and secondarily from net land use change emissions."

3. Humans are the cause and 'fingerprints' on current climate change are proof. "Human influence on the climate system is clear. This is evident from the increasing greenhouse gas concentrations in the atmosphere, positive radiative forcing, observed warming, and understanding of the climate system. This evidence for human influence has grown since the IPCC's Fourth Assessment Report, released in 2007. It is extremely likely that human influence has been the dominant cause of the observed warming since the mid-20th century."

4. Climate change will continue into the future of the planet indefinitely due to the climate system inertia and more greenhouse gases that are already in the system. "Continued emissions of greenhouse gases will cause further warming and changes in all components of the climate system. Limiting climate change will require substantial and sustained reductions of greenhouse gas emissions."

5. There will be more and longer heat waves. "It is very likely that heat waves will occur with a higher frequency and duration."
6. There will be more and bigger storms. "Extreme precipitation events over most of the mid-latitude land masses and over wet tropical regions will very likely become more intense and more frequent by the end of this century, as global mean surface temperature increases."
7. There will be less ice and snow as the Earth warms and the seasons change (summer gets longer and winter shorter). "It is very likely that the Arctic sea ice cover will continue to shrink and thin and that Northern Hemisphere spring snow cover will decrease during the 21st century as global mean surface temperature rises. Global glacier volume will further decrease."
8. There will be higher sea levels as the ice sheets collapse. "Global mean sea level will continue to rise during the 21st century. Under all RCP scenarios the rate of sea level rise will very likely exceed that observed during 1971–2010 due to increased ocean warming and increased loss of mass from glaciers and ice sheets."
9. Ocean acidification will continue. Carbon dioxide and sea water will continue to form acid. "The ocean has absorbed about 30 % of the emitted anthropogenic carbon dioxide; causing ocean acidification…Further uptake of carbon by the ocean will increase ocean acidification."
10. These climate changes will be with us for a long time. "Cumulative emissions of $CO_2$ largely determine global mean surface warming by the late 21st century and beyond. Most aspects of climate change will persist for many centuries even if emissions of $CO_2$ are stopped. This represents a substantial multi-century climate change commitment created by past, present and future emissions of $CO_2$."
11. Humans have to make choices for the future. "Limiting the warming caused by anthropogenic $CO_2$ emissions alone with a probability of >33, >50, and >66 % to <2 °C since the period 1861–1880, will require cumulative $CO_2$ emissions from all anthropogenic sources to stay between 0 and about 1,560 GtC* 0 and about 1,210 GtC, and 0 and about 1,000 GtC since that period respectively. These upper amounts are reduced to about 880, 840, and 800 GtC respectively, when accounting for non-$CO_2$ forcings." *(GtC = gigaton of carbon = 1 billion tons of carbon)
12. Global warming is irreversible. "A large fraction of anthropogenic climate change resulting from $CO_2$ emissions is irreversible on a multi-century to millennial time scale, except in the case of a large net removal of $CO_2$ from the atmosphere over a sustained period."
13. Concentrate on the long-term trends, not short time periods. "Due to natural variability, trends based on short records are very sensitive to the beginning and end dates and do not in general reflect long-term climate trends. As one example, the rate of warming over the past 15 years (1998–2012; 0.05 [−0.05 to +0.15] °C per decade), which begins with a strong El Niño, is smaller than the rate calculated since 1951 (1951–2012; 0.12 [0.08–0.14] °C per decade)."

Because of its scientific and intergovernmental nature, the IPCC has a unique opportunity to provide rigorous and balanced scientific information to decision

**Table 1.1** IPCC AR5 2013 terminology for probability of an event

| Term* | Likelihood of the outcome |
|---|---|
| Virtually certain | 99–100 % probability |
| Very likely | 90–100 % probability |
| Likely | 66–100 % probability |
| About as likely as not | 33–66 % probability |
| Unlikely | 0–33 % probability |
| Very unlikely | 0–10 % probability |
| Exceptionally unlikely | 0–1 % probability |

*Additional terms (extremely likely: 95–100 % probability, more likely than not: >50–100 % probability, and extremely unlikely: 0–5 % probability) may also be used when appropriate

makers. By endorsing the IPCC reports, governments acknowledge the authority of their scientific content. The work of the organization is therefore policy-relevant and yet policy-neutral, never policy-prescriptive; but the IPCC is policy-driven. Without a need for a climate change policy, there would be no need for an IPCC.

The IPCC assessments also include supplementary and special reports. These augment the assessment reports and are also available at the IPCC website given below where more information on the IPCC can be found: http://www.ipcc.ch/organization/organization.shtml#.UovgP0HZhn8.

The IPCC reports are not easy reading for most people, even those with a scientific background. There is so much input from individual scientists, government policymakers, and a variety of others that agreed-upon final language of the reports is sometimes, more often than not, very difficult to read. Their treatment of uncertainty in the AR5 serves as an example.

The terms in Table 1.1 have been used by the IPCC (AR5 2013) to indicate the assessed likelihood of the outcome of an event.

The above terms are used throughout IPCC's Working Group I report *The Physical Science Basis* (2013) and they result in making the text of the report difficult to read and interpret. These designations disrupt the desired flow and continuity of the text. Even with the awkward writing style, the IPCC AR5 *The Physical Science Basis* contains much useful and up-to-date (as of 2012) climate change information.

The IPCC, since its inception, has become extremely important in influencing climate change science, future research, and conclusions reached by climate scientists.

## 1.4 Representative Concentration Pathways

The IPCC AR5 uses "Representative Concentration Pathways" (RCPs) each of which represents an emission scenario; a new set of scenarios that replaces the Special Report on Emissions Scenarios (SRES) standards employed in two of the previous IPCC Assessment Reports. There are four pathways: RCP8.5, RCP6, RCP4.5 and RCP2.6—the last is also referred to as RCP3-PD. (The numbers refer to forcings for each RCP; PD stands for Peak and Decline.)

An excellent and readable guide to RCPs may be found at the following website: http://www.skepticalscience.com/rcp.php.

The purpose of working with scenarios is not to predict the future but to better understand uncertainties and to examine possible alternatives to consider how practical different decisions or options may be under a variety of different possible futures.

Scenarios provide a convenient starting point for additional research on a topic. Climate change is a highly complex subject (or group of subjects) and there are many teams throughout the world working on different aspects of Earth's climate. It is important to have as much standardization as possible with these teams to avoid repetition. If they all used different metrics, made different assumptions about baselines and starting points, then it would be very difficult to compare one study to another. In the same way, models could not be validated against other different, independent models, and communication between climate modeling groups would be made more complex and time-consuming.

Another problem is the cost of running models on mainframe computers. The powerful computers required for climate models are in short supply and great demand. Simulation programming that had to start from scratch for each run would be wholly impractical and costly. Scenarios provide a framework by which the process of building experiments can be streamlined saving time and money.

A RCP scenario consists of a large set of numbers. RCP data is in tables similar to a spreadsheet. For each category of emissions, an RCP contains a set of starting values and the estimated emissions up to 2100, based on assumptions about economic activity, energy sources, population growth and other socio-economic factors.

Modelers download the database sets to initialize their models, which jump-start what would otherwise be a very lengthy process, one that each modeling team would have to attempt, thus duplicating effort. RCPs and previous scenarios were created exactly to avoid such duplication, and the inevitable initialization inconsistencies that would ensue.

Each RCP contains the same categories of data, but the values vary a great deal, reflecting different emission trajectories over time as determined by the underlying socio-economic assumptions (which are unique to each RCP).

Modeling communities made clear their interest in exploring longer-term processes. Each RCP was designed to go to 2100. To facilitate these investigations, a single extension was developed for each RCP, extending the scenarios up to the year 2300. These data form the Extended Concentration Pathways (ECPs) and are explained in Table 1.2.

## 1.5 Radiative Forcing

Radiative forcing (RF) is a measure of the net change in the energy balance of the Earth system in response to some perturbation, with positive RF leading to a warming and negative RF to a cooling.

**Table 1.2**  Parameters, extended concentration pathways (ECPs), and a generic rule for each

| Parameter | RCP/ECP | Generic rule |
|---|---|---|
| $CO_2$ and other well-mixed GHGs | RCP/ECP8.5 | Follow stylized emission trajectory that leads to stabilization at 12 W/m² |
| | RCP/ECP6 | Stabilize concentrations in 2150 (around 6.0 W/m²) |
| | RCP/ECP4.5 | Stabilize concentrations in 2150 (around 4.5 W/m²) |
| | RCP/ECP3-PD | Keep emissions a 2100 level |
| | RCP/SCP6 to 4.5 | Return radiative forcing of all gases from RCP6.0 to RCP4.5 levels by 2250 |
| Reactive gases | All RCP/ECPs | Keep constant at 2100 level |
| | RCP/SCP6 to 4.5 | Scale forcing of reactive gases with GHG forcing |
| Land use | All RCP/ECPs | Keep constant at 2100 level |

A single extension was developed for each RCP, extending the scenarios up to the year 2300. These data form the Extended Concentration Pathways (ECPs) (from van Vuuren et al. 2011)

The main climate forcing agents or drivers are shown in Fig. 1.3. Pay particular attention to the total anthropogenic RF relative to 1,750 and to the individual forcing agents and their impacts on the climate system.

From Fig. 1.3, the radiative forcing from carbon dioxide is the most effective of all the forcings.

## 1.6  Earth's Energy Imbalance and Energy Flow

Earth's energy imbalance is on the order of +0.58 W/m² ±0.15 to 1 W/m² averaged over the whole Earth (Hansen et al. 2011; IPCC 2013). The imbalance fluctuates according to the vicissitudes of output from the Sun and feedbacks. Earth's energy imbalance is the difference between the amount of solar energy coming to the Earth and the amount of energy the planet radiates to space as heat. If the imbalance is positive, more energy coming in than going out, we can expect Earth to become warmer, but cooler if the imbalance is negative. Earth's energy imbalance is thus the single most crucial measure of the status of Earth's climate, whether it is warming or cooling, and it defines expectations for future climate change. In order for the climate to stabilize, the outflow of energy must equal the inflow.

The amount of energy reaching the top of Earth's atmosphere each second on a surface area of one square meter facing the Sun during daytime is about 1,370 W, and the amount of energy per square meter per second averaged over the entire planet is one-quarter of this or 342.5 W/m². See Fig. 1.4 to see what happens to this energy as it flows throughout the climate system.

Energy imbalance arises because of changes in the climate forcings acting on the planet in combination with the planet's thermal inertia and feedbacks. For

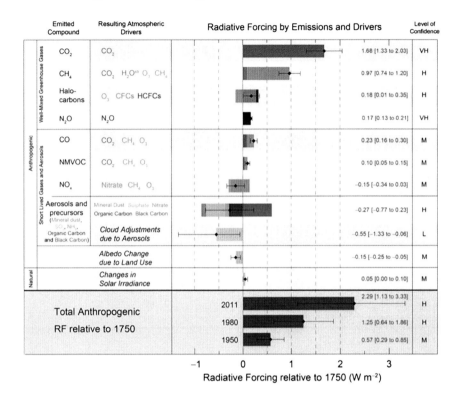

**Fig. 1.3** Radiative forcing estimates in 2011 relative to 1750 and aggregated uncertainties for the main drivers of climate change. Values are global average radiative forcing, partitioned according to the emitted compounds or processes that result in a combination of drivers. The best estimates of the net radiative forcing are shown as *black diamonds* with corresponding uncertainty intervals; the numerical values are provided on the *right of the figure*, together with the confidence level in the net forcing (*VH* very high, *H* high, *M* medium, *L* low, *VL* very low). Albedo forcing due to black carbon on snow and ice is included in the black carbon aerosol bar. Small forcings due to contrails (0.05 W m$^{-2}$, including contrail induced cirrus), and HFCs, PFCs and SF$_6$ (total 0.03 W m$^{-2}$) are not shown. Concentration-based RFs for gases can be obtained by summing the like-shaded bars. Volcanic forcing is not included as its episodic nature makes is difficult to compare to other forcing mechanisms. Total anthropogenic radiative forcing is provided for three different years relative to 1750 (IPCC AR5 2013)

example, if the Sun becomes brighter and more of Sun's energy impacts Earth, that is a positive forcing that will cause warming. Forcings cause Earth's climate to change. Earth's thermal inertia is due mainly to the World Ocean.

Figure 1.4 shows the flow of energy in Watts per square meter (W/m$^2$) throughout the Earth's climate system in a state of equilibrium. Note that the energy leaving the system in the Fig. 1.4 equals that coming into the system (342 W/m$^2$). Also note that 324 W/m$^2$ is back radiation from greenhouse gases (GHGs). The GHGs trap heat energy and radiate it back to the surface; also note the thermals and evapotranspiration.

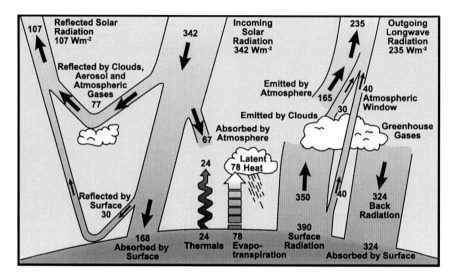

**Fig. 1.4** Estimate of the Earth's annual and global mean energy balance. Over the long term, the amount of incoming solar radiation absorbed by the Earth and atmosphere is balanced by the Earth and atmosphere releasing the same amount of outgoing longwave radiation. About half of the incoming solar radiation is absorbed by the Earth's surface. This energy is transferred to the atmosphere by warming the air in contact with the surface (thermals), by evapotranspiration and by longwave radiation that is absorbed by clouds and greenhouse gases. The atmosphere in turn radiates longwave energy back to Earth as well as out to space (*Source* Kiehl and Trenberth (1997), from IPCC AR4 2007b)

As can be seen from Fig. 1.4, 324 W/m$^2$ of Earth's energy budget gets reflected back toward Earth's surface causing what is known as the "greenhouse effect." The enhanced greenhouse effect (due to greenhouse gases added by humans) and the constant stream of new energy coming into the climate system from the Sun are causing Earth to warm rapidly compared to warming of the past that happened over millennia.

## 1.7 Rising Temperature

Although it was cold in parts of the world (especially in the U.S.), globally November 2013 was the hottest November since records began in 1880. Australia recently reported that 2013 was their hottest year since record keeping began in the early 1900s. Earth is still warming and the main causes of this warming are (1) the positive energy imbalance and (2) anthropogenic carbon dioxide.

Since 1880 Earth's global temperature has increased about 0.8–1.0 °C (the range of temperatures exist because of the uncertainty of 19th and early 20th century data). The temperature rise during this period is shown in Fig. 1.5.

The current degree of imbalance is sustained by the enhanced greenhouse effect. The enhanced greenhouse effect is caused primarily by the emissions from burning

**Fig. 1.5** Line plot of global
mean land-ocean temperature
index, 1880–2014, with the
base period 1951–1980. The
*black line* with the *squares*
is the annual mean and the
*solid line* is the 5-year mean.
The *vertical bars* show
uncertainty estimates (from
NASA/GISS, Public Domain)

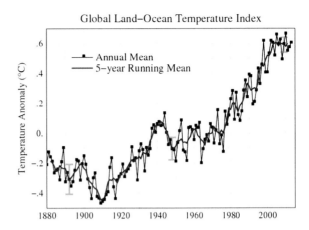

fossil fuels, most notably carbon dioxide from coal-burning power plants, automobile exhausts, the making of cement, changes in land use (especially deforestation), and sources of other greenhouse gases. These causes are all due to humankind's impact on Planet Earth. Prior to human influence on Earth's climate, the temperature and greenhouse gases concentration had remained fairly steady for at least 11,300 years (Marcott et al. 2013).

One degree Celsius may not sound like much. When it is already having extreme effects on weather, astute people are becoming very concerned. Familiar weather patterns are being disrupted, polar ice sheets are melting, sea ice around the North Pole is disappearing, storms are becoming more intense, tropical diseases are spreading into higher latitudes and altitudes, and seasons of the year are changing. All of these would be reasons for concern even if the planet was not warming, and all because of 1 °C. What will happen with 2, 3, 4, 5, 6, or 10 °C?

The 20 warmest years ever recorded by humankind have all occurred since 1981 and the 10 warmest have all occurred in the last 12 years (from this writing in February 2014). The globally-averaged temperature for October 2013 was the seventh warmest October since record keeping began in 1880. October 2013 also marks the 37th consecutive October and 344th consecutive month with a global temperature above the 20th century average (NCDC, October 2013). November 2013 was globally the warmest November in recorded history.

The temperatures in the graph (Fig. 1.5) are given as anomalies. The method of graphing temperatures as anomalies relative to a time interval or base period (in the graph (Fig. 1.5), the interval used for comparison is the mean temperature for the years 1951–1980) is practical because actual temperature data would be meaningless. One would not be able to compare temperature data in a meaningful way for several reasons; for example, if one reading was taken in a valley at night and the next taken during the day on a mountain top, or if two readings were taken at the same location but one at noon and the other at midnight.

Marcott et al. (2013) showed global temperature rose faster in the past century than it has since the end of the last ice age, more than 11,300 years ago. Using fossils, corals, ice cores and tree rings, a study in the journal *Science* in March 2013 became the first to take an 11,300-year look back in time to determine Earth's temperature history.

Deniers of climate change/global warming say that warming stopped in 1998. First, 16 years of data does not meet the World Meteorological Organization's criterion for a legitimate climate data set. The WMO defines at least a 30-year period for establishing a climate trend. Second, as one can see from the above illustration, there has been a leveling out and slight decrease in warming since about 1998. However, this does not mean that the warming has stopped. Notice the trends in the above graph. According to the denier's logic, warming also stopped in 1900, 1940, 1960, etc. However, the overall trend is what is important and not 'cherry picking' a few dates from this trend. Several recent research papers show that the climate system is warming at even a faster rate over the past 30 years with about 93.4 % of the warming going into the ocean, and the ocean is warming at depth.

## 1.8 Solar Irradiance

The vast majority of Earth's energy comes from the Sun. It's this solar energy that drives the climate system. Solar variations, together with volcanic activity, are hypothesized to have contributed to climate change; for example, during the Maunder Minimum (an episode of low sunspot activity that occurred from approximately 1,645 to 1,715). Changes in solar brightness, however, are too weak to explain the recent consistent global warming.

Figure 1.6 shows comparisons with daily-averaged values of the Sun's total irradiance (TSI) from radiometers on different space platforms.

Figure 1.7 from NASA shows solar irradiance outside the atmosphere and solar irradiance at sea level. Ozone ($O_3$), carbon dioxide ($CO_2$), and water vapor ($H_2O$)

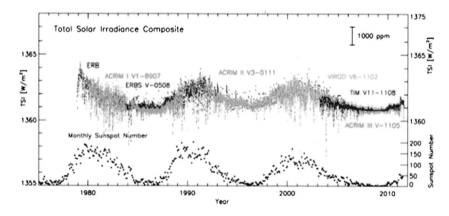

**Fig. 1.6** Space-borne measurements of the total solar irradiance (TSI) show ~0.1 % variations with solar activity on 11-year and shorter timescales. These data have been corrected for calibration offsets between the various instruments used to measure TSI. *Source* Courtesy of Greg Kopp, University of Colorado (from NASA, Public Domain)

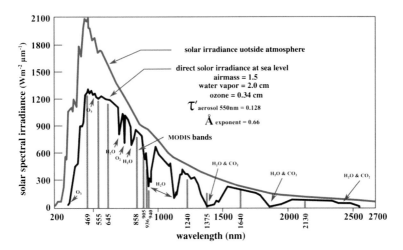

**Fig. 1.7** Solar irradiance spectrum. Moderate Resolution Imaging Spectroradiometer (MODIS) bands shown (*orange vertical lines*). Solar spectral irradiance incident on the top of the atmosphere (*green or outer curve*) and transmitted through the atmosphere to the Earth's surface (*brown or inner curve*). Major absorption bands in the atmosphere are clearly apparent (NASA, Public Domain)

absorption bands are clearly visible at their wavelengths. These three-part molecules vibrate in wavelengths similar to those of longwave radiation from Earth's surface. As such, they trap heat at those wavelengths and re-radiate this heat in all directions, some of it back to Earth's surface. As these molecules trap heat, they also warm. The continuous warming of the troposphere causes that layer of the atmosphere to rise, forcing the tropopause upward.

## 1.9   Carbon Dioxide's Role in the Greenhouse Effect

Carbon dioxide ($CO_2$) is a minor constituent of Earth's atmosphere. Carbon dioxide is only 0.0400 % of the atmosphere. How could it be so important in global warming studies? By the time you finish this section, you will know the answer to that question.

The greenhouse effect is due to Earth's atmosphere (see color or shaded bands in Fig. 1.8) that traps heat radiation from Earth's surface. The greenhouse effect has been enhanced by greenhouse gases that have been added by humans that keep heat from radiating back to space. Because of the 'enhanced greenhouse effect,' the Earth will continue to get warmer until the energy imbalance reaches a state of balance (inflow equals outflow).

Earth's atmosphere is able to retain heat because of the properties of greenhouse gases (GHGs) that trap heat radiating from Earth's surface. The GHGs allow Sunlight (ultraviolet radiation or UV) to pass through them but trap heat energy (infrared radiation or IR) coming from Earth's surface. These greenhouse gas molecules vibrate in wavelengths that capture heat and send (radiate) it in all

**Fig. 1.8** Though astronauts and cosmonauts often encounter striking scenes of Earth's limb, this unique image, part of a series over Earth's colorful horizon, has the added feature of a silhouette of the space shuttle Endeavour. The image was photographed by an Expedition 22 crew member prior to STS-130 rendezvous and docking operations with the International Space Station. The *orange layer* closer to the dark Earth is the troposphere, where all of the weather and clouds which we typically watch and experience are generated and contained. This *orange layer* gives way to the whitish Stratosphere and then into the Mesosphere. In some frames the *black color* is part of a window frame rather than the blackness of space (NASA, Public Domain)

directions including back in the direction of the Earth's surface. The molecular structure of $CO_2$ is such that it is "tuned" to the wavelengths of infrared (heat) radiation emitted by the Earth's surface, in particular to the 15 $\mu$m band. The molecules resonate, their vibrations absorbing the energy of the infrared radiation. It is vibrating molecules that give us the sensation of heat, and it is by this mechanism that heat energy is trapped by the atmosphere and re-radiated to Earth's surface.

Each of the most abundant greenhouse gases consists of two elements (with one exception, ozone, $O_3$), the most common and effective of these being water vapor ($H_2O$). Although water vapor is the most common greenhouse gas, it doesn't stay in the atmosphere long, is not well mixed in the atmosphere, and it is constantly being flushed out falling to Earth's surface as precipitation.

Other greenhouse gases are carbon dioxide ($CO_2$), methane ($CH_4$), and nitrous oxide ($N_2O$) all of which have similar properties. The second most important greenhouse gas is carbon dioxide*. Carbon dioxide concentration in the atmosphere has been increasing at least since the Industrial Revolution and probably much earlier, when humans started to clear land of forests by using fire. The greatest single source of carbon dioxide today is the burning of fossil fuels, and more specifically, burning coal to generate electricity.

## 1.10  Carbon Dioxide and Carbon

Carbon and carbon dioxide are not the same. Both terms are often used as if they were synonymous, but they are not. Carbon (C) is an element and carbon dioxide ($CO_2$) is a compound. The fraction of carbon in carbon dioxide is the ratio of their atomic weights. The atomic weight of the vast majority of carbon is 12 atomic mass units, while the atomic weight of carbon dioxide is 44, because it includes two oxygen atoms that each has an atomic weight of 16. So, to switch from one to the other, use the formula: One unit of carbon equals 44/12 = 11/3 = 3.67 units of carbon dioxide. Thus 11 tons of carbon dioxide equals 3 tons of carbon. The ratio of carbon dioxide to carbon is 11:3 or a ratio of carbon dioxide to carbon is 3.67–1. This relationship is somewhat confusing because carbon is a solid in its native state and carbon dioxide is a gas. It would seem that carbon would be heavier than carbon dioxide because solids are heavier than gases. But the reverse is true due to the atomic weights.

There are many references to carbon and carbon dioxide in this book, and in climate science, and the reader should understand the difference between them. As a matter of convenience, equivalences of carbon and carbon dioxide terms are given in parentheses; (*1 Petagram of carbon = 1 Gigaton of carbon = 1 GtC. This corresponds to 3.67 GtCO$_2$.).

Earth's climate system's energy imbalance is of +0.58 to +1.0 W/m$^2$ (Trenberth et al. 2009; Hansen et al. 2011; IPCC AR5 2013). Carbon dioxide levels have continued to climb (see Fig. 1.9). So where is this warming going? It doesn't show up just above Earth's surface on land or in shallow marine layers. Surface ocean and land temperatures seem to have stabilized since 1998; but global warming affects the whole Earth, not just the bottom of the atmosphere.

The warming shows up in deeper marine waters. Climate scientists have known for some time that the large majority of warming goes into the ocean (93.4 %). More than 30 % of the heat was deposited in the ocean below 700 m in the post-2000 ocean reanalysis system and was identified mainly with changes in the

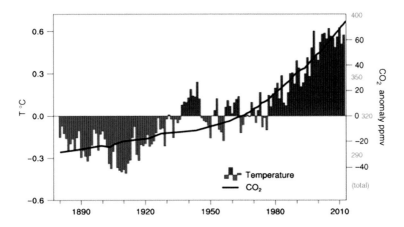

**Fig. 1.9** Estimated changes in annual global mean surface temperatures (°C) and CO$_2$ concentrations (*thick black line*) since 1880. The changes are shown as differences (anomalies) from the 1901 to 2000 average values. Carbon dioxide concentrations since 1957 are from direct measurements at Mauna Loa, Hawaii, whereas earlier estimates are derived from ice core records. The scale for CO$_2$ concentrations is in parts per million (ppm) by volume, relative to a mean of 320 ppm, whereas the temperature anomalies are relative to a mean of 13.9 °C (57 °F) (from Kevin E. Trenberth, John T. Fasullo, Article first published online: 5 Dec 2013, doi:10.1002/2013EF000165)

tropical and subtropical winds in the Pacific. The increased heating below 700 m in the World Ocean of about 0.2 $W/m^2$ globally is revealed after year 2000 in the reanalysis by NOAA (Trenberth and Fasullo 2013).

### 1.10.1  The Enhanced Greenhouse Effect

The enhanced greenhouse effect is the effect that humans have had and are having on the normal greenhouse effect. This warming has already resulted in a 0.8–1 °C temperature rise since the Industrial Revolution (actually since 1880) and the Earth's temperature continues to rise today at a steadily increasing rate.

There is agreement among European nations, the EU, and some scientists that the total of Earth's global temperature rise should be kept below 2 °C, from the temperature before the Industrial Revolution. We have already allowed a 1 °C rise so we have only 1 °C remaining.

The European Union (EU) has set 2 °C as the limit that is acceptable. In order to keep the temperature below 2 °C, calculations show that carbon dioxide should not exceed 350 parts per million (ppm) in the atmosphere. Carbon dioxide has already risen above 400 ppm. This would mean that humans must develop a method of removing carbon dioxide from the atmosphere to reduce its concentration back to 350 ppm. This would also mean that no more $CO_2$ can be added to the atmosphere. This is not going to happen in the near future unless governments of the world understand that a global agreement restricting $CO_2$ emissions is needed and arrangements are made to enforce it.

### 1.10.2  Climate Sensitivity

Climate sensitivity is the equilibrium temperature change in response to changes in the radiative forcing. How sensitive is Earth's climate to a change, say a doubling of carbon dioxide from pre-industrial levels? The pre-industrial level of $CO_2$ in the atmosphere is estimated to have been 280 ppm. A doubling of $CO_2$ would then be 560 ppm. How much would the global average temperature rise if the $CO_2$ concentration in the atmosphere doubled to 560 ppm? The amount of temperature rise in degrees Celsius with a doubling of $CO_2$ is commonly how the climate's sensitivity is expressed.

The IPCC (IPCC AR5 2013) estimates 1.5–4.5 °C to be the likely range of climate sensitivity to a doubling of carbon dioxide in the atmosphere from 280 to 560 ppm. The carbon dioxide in the atmosphere has already increased 120 ppm from the pre-industrial level (to 400 from 280 ppm) so there is a total increase of 160 ppm to go before we see the doubling of $CO_2$ and a temperature increase of perhaps 4.5 °C. Earth has already warmed 0.8–1 °C since the Industrial Revolution, so we can warm an additional 3.5 °C when we reach 560 ppm of

$CO_2$? But what if the climate sensitivity is greater than 4.5 °C and instead of doubling the $CO_2$ we triple it? We could be wrong, and the climate sensitivity could be greater than 4.5 °C. What if we allow $CO_2$ to double or triple or quadruple and the global temperature increases 6, 8, 10 or even 12 °C? Humans could not survive at 6 °C above our current average global temperature and perhaps not at 4.5 °C.

A global temperature increase of 4.5 °C would most likely cause all ice on Earth to melt and sea level to rise around the world about 260 ft. If all the ice melts, all coastal and inland cities below 260 ft in elevation will be inundated and their populations displaced to higher elevations. A world 4.5 °C warmer than the pre-industrial level would be a world in which humankind has never experienced. Heat waves would be brutal and last longer. More forests would burn. More water would evaporate from the barren ground, lakes and rivers would dry, and drinking water would become scarce.

The IPCC AR5 Report (30 September 2013), *The Physical Science Basis* uses Equilibrium Climate Sensitivity (ECS) to mean how much warming will take place for the climate system to reach a state of equilibrium from a given forcing. Estimates of the Equilibrium Climate Sensitivity, based on observed climate change, climate models, and feedback analysis as well as paleoclimate evidence, indicate that ECS is positive, likely in the range 1.5–4.5 °C with high confidence, extremely unlikely less than 1 °C (high confidence) and very unlikely greater than 6 °C (medium confidence). Earth system sensitivity over thousands-of-year time-scales including long-term feedbacks not typically included in models could be significantly higher than the ECS.

## *1.10.3  Carbon Capture and Sequestration*

Carbon capture and sequestration (or carbon capture and storage; CCS) is a misuse of the term "carbon." What is meant is carbon dioxide. CCS is the process of capturing waste carbon dioxide from large point sources, such as fossil fuel power plants, transporting it to a storage site if necessary, and depositing it where it will not enter the atmosphere. Normally this would be an underground geological formation.

Dr. James Hansen, formerly the Director of NASA/GISS and now an Adjunct Professor at Columbia University and one of the world's most esteemed climate scientists, has stated that we should reduce $CO_2$ concentration in the atmosphere to 350 ppm to avoid catastrophic changes to Earth's climate and to humankind. In order to return to 350 ppm $CO_2$ in the atmosphere it will be necessary to reduce the carbon dioxide concentration by at least 50 ppm. It will be necessary to remove the 50 ppm and then keep it at or below 350 ppm *ad infinitum*.

It is much better and less expensive to not add carbon dioxide in the first place, but to bring its level back to 350 ppm from its current level of 400 ppm means extracting 50 ppm of carbon dioxide already in the atmosphere and hiding it away permanently. After removal of the $CO_2$, to keep the level at 350 ppm it will be necessary to continue to remove the $CO_2$ at the rate it is being added to the atmosphere, or at a greater rate when it exceeds 350 ppm to bring it below 350 ppm.

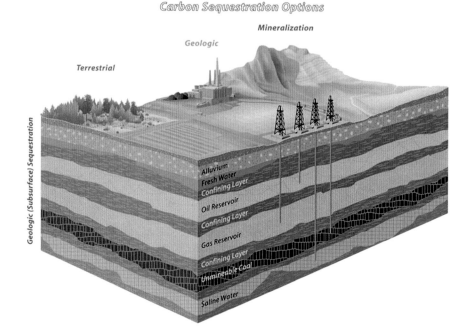

**Fig. 1.10** Schematic showing both terrestrial and geological sequestration of carbon dioxide emissions from a coal-fired plant (*Credit* Genevieve Young, Colorado Geological Survey)

Carbon dioxide capture and sequestration (CCS) may be an extremely expensive and impractical way of dealing with the problem, but there are several studies underway. Carbon dioxide is captured at the source, compressed, and then transported to a sequestration or storage facility by pipeline deep underground or by truck or rail to the point where it is to be sequestered. If the carbon dioxide is to be sequestered deep underground, the geologic conditions must be such that none of the gas will escape. There are several sites worldwide being tested for their adequacy for sequestration.

The United States has at least 2,400 billion metric tons of possible carbon dioxide ($CO_2$) storage resource in saline formations, oil and gas reservoirs, and unmineable coal seams. Other geologic conditions are being evaluated by the U.S. Department of Energy and the U.S. Geological Survey and perhaps others. A diagrammatic illustration of possible CCS is given in Fig. 1.10.

### 1.10.4  Tipping Points

A tipping point in climate change science is a point, when reached, is set in motion a process that cannot be stopped. For example, a tipping point will be reached when the temperature is great enough to melt all ice on Earth. Both the

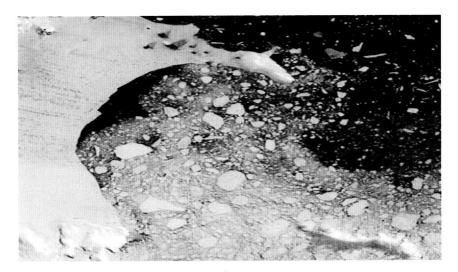

**Fig. 1.11** Larsen B ice shelf collapse: Abrupt climate change—large shifts in climate that take place within decades or even years—is the topic of increasing scientific research because of the potential for such changes to happen faster than society or ecosystems could adapt (*Credit* Landsat 7 Science Team and NASA GSFC, Public Domain)

Greenland and Antarctic ice sheets already show signs of degradation (Fig. 1.11), and Arctic sea ice is disappearing rapidly.

A recent (2013) report from the U.S. National Research Council elaborates on the idea of abrupt climate change, stating that even steady, gradual change in the physical climate system can have abrupt impacts elsewhere, in human infrastructure and ecosystems for example, if critical thresholds (tipping points) are crossed.

Earth's temperature may already be past the tipping point for melting ice on Earth. An example of Greenland's ice sheet melting is shown in Fig. 1.12.

These changes in the stability of the two ice sheets (Greenland and Antarctica) are of major significance in determining the future of humankind on the planet. The Greenland and Antarctic ice sheets are massive reservoirs of fresh water and have a direct effect on sea level. If there is a sudden rise in sea level, millions of people will be forced to move inland. This will cause severe overcrowding in inland communities that are unprepared for such an influx of people. As sea level continues to rise, more people will have to move inland and population pressures will become intense.

Tipping points may be reached suddenly or gradually over decades, hundreds, or even thousands of years and once the point is reached it can have disastrous results. For example, a plant species that when processed contains a life-saving drug, may have been living in a region that is gradually increasing in temperature and suddenly, the limit of tolerance for temperature of the species is reached. The species becomes extinct very rapidly once the tipping point is reached and the source for the life-saving drug is gone forever. This is surely happening today to some plants as well as to some animals as their limits of tolerance to temperature and other environmental stresses are being exceeded.

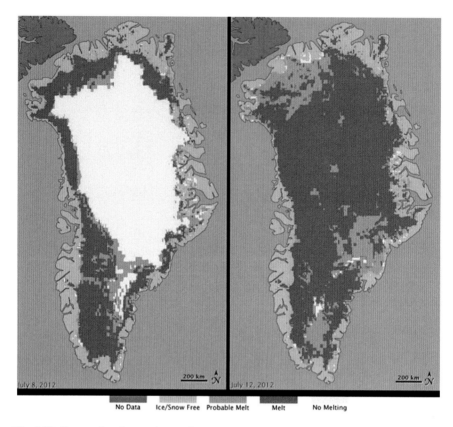

**Fig. 1.12** Extent of surface melt over Greenland's ice sheet on July 8 (*left*) and July 12 (*right*), 2012. Measurements from three satellites showed that on July 8, about 40 % of the ice sheet had undergone thawing at or near the surface. In just a few days, the melting had dramatically accelerated and an estimated 97 % of the ice sheet surface had thawed by July 12. In the image, the areas classified as "probable melt" (*lighter*) correspond to those sites where at least one satellite detected surface melting. The areas classified as "melt" (*darker*) correspond to sites where two or three satellites detected surface melting. The satellites are measuring different physical properties at different scales and are passing over Greenland at different times. As a whole, they provide a picture of an extreme melt event about which scientists are very confident (*Credit* Nicolo E. DiGirolamo, SSAI/NASA GSFC, and Jesse Allen, NASA Earth Observatory, NASA, Public Domain)

## 1.11 Climate Forcing, and Climate Feedbacks

Two very important concepts in climate change science are those of climate forcing and climate feedbacks. Climate forcing causes the climate system to change. Feedbacks either enhance or retard a forcing. There are positive and negative forcings just as there are positive and negative feedbacks.

### 1.11.1  Climate Forcing

Radiative forcing was introduced in Sect. 1.4 and Fig. 1.3. Climate systems remain stable until there is a perturbation or force that changes the system. These perturbations are called forcings because they force the climate system to change or react to the forcing. The forcing may be either positive (causing the climate system to warm) or negative (causing the system to cool).

An example of a positive climate forcing is when the Sun increases the solar energy it emits. If the temperature of Earth is warming, then an increase in solar energy results in an additional temperature increase on Earth. The climate system is forced to continue to warm, perhaps at a greater rate.

An example of a negative forcing is when the Sun goes into a minimum (energy output is less), the Earth's climate cools.

Radiative forcing has to do with energy impacting the Earth either from the Sun (solar radiation) or from within the Earth's climate system, such as the buildup of greenhouse gases in the atmosphere; or volcanoes. Volcanoes spewing fine particles of ash high in the atmosphere cause a negative forcing by blocking a portion of the Sun's rays from hitting Earth's surface and the Earth's global average temperature will fall. If the forcing is a pulse of new energy that causes a temperature change, the change will continue until a new stable temperature is attained.

Earth's climate system reacts to a forcing by attempting to return to a state of equilibrium. Attaining a new state of equilibrium usually takes a long time due to the inertia in the system. Climate system inertia is due largely to the slow reaction time of the Earth's World Ocean.

### 1.11.2  Climate Feedbacks

Climate feedbacks either enhance a climate forcing (positive feedback) or retard a climate forcing (negative feedback).

An example of a positive feedback is when warming is enhanced by something within the climate system that increases the warming, such as melting glaciers, sea ice, and ice sheets. As glaciers, sea ice, and ice sheets melt, they allow more of the Sun's radiation to directly impact the surface of the Earth and further warm the surface. In this case, a feedback loop is formed as more ice loss causes more warming which causes more ice loss, and so on.

An example of a negative feedback is when a warming Earth experiences a volcanic eruption (or series of eruptions) that causes a cloud of ash to find its way into the upper atmosphere and block a portion of the Sun's rays from striking Earth's surface. The Earth will cool.

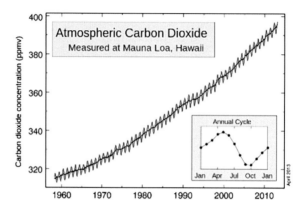

**Fig. 1.13** The carbon dioxide data, measured as the mole fraction in dry air, on Mauna Loa constitute the longest record of direct measurements of $CO_2$ in the atmosphere. They were started by C. David Keeling of the Scripps Institution of Oceanography in March of 1958 at a facility that was taken over by the National Oceanic and Atmospheric Administration. The *black line* represents the seasonally corrected data (NOAA, Public Domain)

### *1.11.3 Charles David Keeling and the Keeling Curve*

Increasing carbon dioxide in the atmosphere is forcing the climate system to warm. No one had accurately measured carbon dioxide in the atmosphere and no time series existed for the chemical's concentration prior to 1958. In 1958, a young scientist by the name of Charles David Keeling began to very accurately measure the carbon dioxide in the atmosphere by using an instrument he had designed and from an observatory he had established on a flank of Mauna Loa in Hawaii. This monitoring of the concentration of $CO_2$ continues today at the Mauna Loa Observatory maintained by NOAA and has resulted in a continuous plot called the Keeling Curve shown in Fig. 1.13.

The Keeling Curve clearly shows the seasonal variation in the concentration of $CO_2$. The concentration is greatest in the winter when photosynthesis is less and when less vegetation is growing in the mid-latitudes. The concentration decreases during summer months when vegetation is using more carbon dioxide in photosynthesis.

The best available evidence suggests the amount of $CO_2$ currently in the air has not been this high (400 ppm) for at least three million years, before humans (*Homo sapiens*) evolved. Most climate scientists believe the rapid $CO_2$ rise portends large changes in the climate and the level of the sea.

## 1.12 Climate Models

Models are representations of things. They may be models of cars, airplanes, buildings or any of a myriad of other things. Most climate models are a complex of differential equations that model Earth's climate, but they range in complexity to models that can be run on an Excel spreadsheet on a laptop to those that have to be run on supercomputers [General Circulation Models or General Climate Models (GCMs)] at special facilities throughout the world that are equipped to run these models.

The models used in climate research range from simple energy balance models (EBMs) to complex Earth System Models (ESMs), the latter requiring state of the art high-performance computing. The choice of model depends directly on the scientific question being addressed. Applications include simulating paleoclimate or historical climate, sensitivity, and process studies for attribution (causation) and physical understanding, predicting near-term climate variability and change on seasonal to decadal time scales, making projections of future climate change over the coming century or more, and down-scaling such projections to provide more detail at the regional and local scale.

Computational cost is a factor in using most of these models and time is expensive using the mainframe resources that are in high demand. Simplified models (with reduced complexity or spatial resolution) can be used when larger ensembles or longer integrations are required. Simplified examples include exploration of parameter sensitivity (parameterization) or simulations of climate change on the millennial or longer time scale.

Figure 1.14 diagrammatically shows the evolution of the complexity of climate modeling from the 1970s through the IPCC AR4 (2007b).

A Global Circulation Model or Global Climate Model (GCM) is a complex mathematical representation of the major climate system components (atmosphere, land surface, ocean surface, ocean depth, glaciers, and sea ice), and their interactions. Earth's energy exchanges between the climate system components are essential to long-term climate projections. The main climate system components treated in a GCM are each of the following:

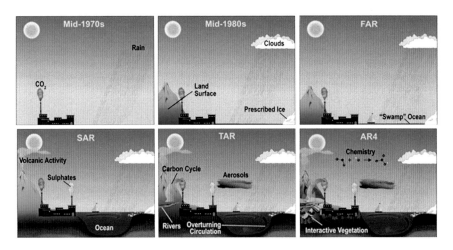

**Fig. 1.14** Evolution of climate models used by the IPCC through AR4; *FAR* first assessment report; *SAR* second assessment report; *TAR* third assessment report; *AR4* fourth assessment report (from the IPCC AR4 2007b)

- The atmospheric component, which simulates clouds and aerosols, and plays a large role in transport of heat and water around the globe;
- The land surface component, which simulates surface characteristics such as vegetation, snow cover, soil, water, rivers, and carbon storing;
- The ocean component, which simulates current movement and mixing, and bio-geochemistry, since the ocean is the dominant reservoir of heat and carbon in the climate system; and
- The glacier and sea ice component, which modulates solar radiation absorption and air-sea heat and water exchanges.

## 1.13  Earth's Atmosphere

The Earth's atmosphere is a surprisingly narrow band of gases, liquids, and solids that extends above our heads for about 600 km as shown in Fig. 1.15.

The vast majority of Earth's weather occurs in the troposphere, that part of the atmosphere in contact with Earth's surface and extending upward to the tropopause. (Not all atmospheric zones mentioned in the text are shown in Fig. 1.15, e.g., the tropopause.)

Above the tropopause is the stratosphere, the lower part of which contains the ozone layer. The stratosphere continues upward to the stratopause. Above the

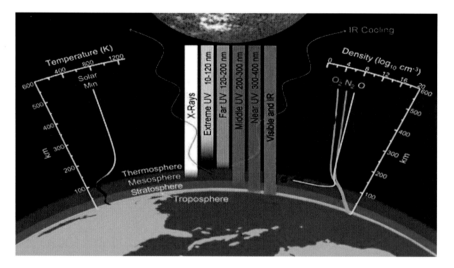

**Fig. 1.15**  The thin envelope of air that surrounds our planet is a mixture of gases, solids and liquids. The mixture of chemicals is not evenly divided. Two elements, nitrogen and oxygen, make up 99 % of the volume of air. The other 1 % is composed of "trace" gases, the most prevalent of which is the inert gaseous element argon. The rest of the trace gases, although present in only minute amounts, are very important to life on earth. Two in particular, carbon dioxide and ozone, can have a large impact on atmospheric processes (from NASA, Public Domain)

stratopause is the mesosphere which extends upward to the mesopause. Above the mesopause is the thermosphere which extends upward to the thermopause. Beyond the thermopause is the exosphere which extends to outer space (and is not labeled on Fig. 1.15).

When considering global warming, it is often thought to be just the atmosphere that is warming. However, it is the whole Earth's surface that is warming including the land surface, the ocean waters' surface, and ice surfaces as well as the atmosphere above these surfaces. Warming is also happening at depth below the land surface as cores from drilling have revealed, and at depth in the deep waters of the ocean basins (to at least a depth of 2,000 m). Ocean waters constitute about 71 % of Earth's surface and direct measurements show that heat is being stored in the deep ocean basins.

## 1.14  Land Surface

Earth's land surface, including that covered by ice, represents only 29 % of Earth's surface area. The land surface can be characterized as consisting of the Earth's surface that is not ocean, is in direct contact with the atmosphere, and consists of a variety of mountains, valleys, plateaus, coastal areas, swamps, lakes, streams, sea ice, ice sheets, and glaciers. It is the land surface that is part of the Geosphere. The rest of the Geosphere lies below the surface.

Land areas heat more rapidly than ocean waters, as anyone swimming in the ocean after sunning at a beach can attest. Land surface temperature is not the same as the air temperature that is included in the daily weather report. In Fig. 1.16, temperatures across land areas of the Earth range from −25 °C (dark) to 45 °C (light). At mid-to-high latitudes, land surface temperatures can vary throughout the year, but equatorial regions tend to remain consistently warm, and Antarctica and Greenland remain consistently cold. Altitude plays a clear role in temperatures, with mountain ranges like the North American Rockies and the Himalayas cooler than other areas at lower elevations at the same latitude.

The vast majority of land on Earth is in the Northern Hemisphere, parts of which are warming faster than that of other parts of the globe.

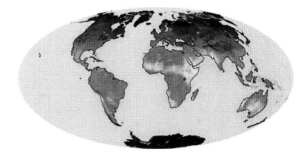

**Fig. 1.16**  Temperature of the land surface. Darker shades are cooler (from NASA, Public Domain)

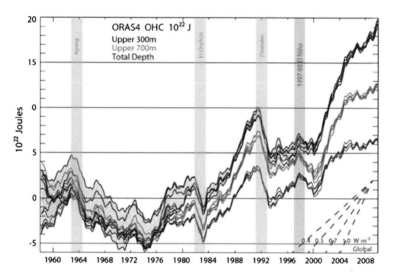

**Fig. 1.17** Amount of heat stored in the whole ocean in past five decades (*top curve*), the top 700 m (*middle curve*) and just the top 300 m (*bottom curve*). *Source* Balmaseda et al. (2013)

## 1.15 The World Ocean

Ocean waters are usually divided into different ocean basins such as the Atlantic, Pacific, Indian, Arctic and Antarctic Ocean. However, all of these oceans are interconnected and waters flow between them, so we can think of them as forming one World Ocean.

The World Ocean is warming slowly compared to land areas of Earth and the atmosphere, but there are recent measurements that show a large portion of recent global warming is finding its way into deeper waters (see Fig. 1.17). A recent study (Balmaseda et al. 2013) found that the oceans are currently absorbing heat 15 times faster than they have at any point during the past 10,000 years and much of this heat is going to deeper waters.

There is a world-wide system of over 3,600 devices called 'Argo floats' that measure ocean temperatures to depths of 2,000 + meters. These floats are pre-programmed to dive, do continuous monitoring, and come back to the surface to transmit their data.

## 1.16 Ocean-Atmosphere Interface

Ocean waters intersect and interact with the atmosphere at what has been referred to as the ocean's 'skin layer.' Despite being only 0.1–1 mm thick on average, this 'skin layer' is the major player in the long-term warming of the ocean. It is not intuitively obvious just how the ocean waters warm. The skin is usually 1°–2° warmer than the atmosphere above it. Heat only travels from higher to lower temperatures so there is a lot of heat escaping from these upper layers of the ocean.

Reducing the heat lost to the atmosphere allows the ocean to steadily warm over time as has been observed over the last half century. The rate of flow of heat out of the ocean is determined by the temperature gradient in the cool 'skin layer' which resides within the thin surface layer of ocean that is in contact with the atmosphere.

The "skin layer" breaks down due to the turbulent action of waves when the upper meter or so of the ocean is in contact with the atmosphere.

To maintain an approximate steady state climate system the ocean and atmosphere must move excess heat from the tropics to the heat-deficit Polar Regions. Additionally, the ocean and atmosphere must move freshwater to balance regions with excess dryness with those of excess rainfall. The movement of freshwater in its vapor, liquid, and solid state is referred to as the hydrological cycle.

In low latitudes the ocean moves more heat poleward than does the atmosphere, but at higher latitudes the atmosphere becomes the big carrier. The wind-driven ocean circulation moves heat mainly on the horizontal plane. For example, in the North Atlantic, warm surface water moves northward within the Gulf Stream on the western side of the ocean, to be balanced by cold surface water moving southward within the Canary Current on the eastern side of the ocean. The thermohaline circulation moves heat mainly in the vertical plane. For example, North Atlantic Deep Water (NADW) with a temperature of about 2 °C flows toward the south in the depth range 2,000–4,000 m to be balanced by warmer water (greater than 4 °C) flowing northward within the upper 1,000 m.

Across the sea surface pass heat, water, momentum, gases, and other materials. Much of the direct and diffuse solar shortwave electromagnetic radiation that reaches the sea surface penetrates the ocean (the ocean has a low albedo compared to clean sea ice), heating the sea water down to about 100–200 m, depending on the water clarity. It is within this thin sunlit surface layer of the ocean that the process of photosynthesis can occur and it is here that the majority of oceanic plants and animals live (as well as the single-celled Protista).

The ocean transmits radiation (IR radiation, i.e., heat) into the atmosphere at much longer wavelengths than that of the solar radiation (UV). The infrared (IR) radiation emitted from the ocean is quickly absorbed and re-emitted by water vapor and carbon dioxide and other greenhouse gases in the lower atmosphere (the troposphere). Much of the radiation from the atmospheric gases, also in the infrared range, is transmitted back to the ocean, reducing the net longwave radiation heat loss of the ocean. The warmer the ocean, the warmer the air above it, increasing the air's capacity to hold water vapor and thus increasing the greenhouse effect. It is very difficult for the ocean to transmit heat by longwave radiation into the atmosphere; the greenhouse gases just re-radiate it back, notably water vapor whose concentration is proportional to the air temperature.

When air is in contact with the ocean it is at a different temperature than that of the sea surface (SST) and heat transfer by conduction takes place. On average the ocean is about 1° or 2° warmer than the atmosphere, so on average, ocean heat is transferred from ocean to atmosphere. The temperature difference causes convection currents to form with warmer air rising and cooler air descending. The heated air is more buoyant than the air above it, so it causes the ocean heat to rise. If the ocean was colder than the atmosphere, the air in contact with the ocean cools, becoming denser. As such, the conduction process does a poor job of carrying the atmosphere's heat into the cool ocean. This occurs over the subtropical upwelling regions of the ocean. The transfer of heat between ocean and atmosphere by conduction is more efficient when the ocean is warmer than the air it is in contact with.

The annual heat flux between ocean and atmosphere is formed by the sum of all of the heat transfer process, solar and terrestrial radiation, heat conduction, and evaporation. While the ocean gains heat in low latitudes and loses heat in high latitudes, the largest heat loss is from the warm Gulf Stream waters off the east coast of the U.S. during the winter, when cold dry continental air spreads over the ocean. An equivalent pattern is found near Japan. It is in these regions that the atmosphere takes over as the major meridional (i.e., longitudinal) heat transfer agent.

Generally, heat transport across latitudes is from the tropics to the Polar Regions, but in the South Atlantic Ocean the oceanic heat transport is directed towards the equator. This is due to the thermohaline circulation as the warm upper kilometer of water is carried northward, across the equator, offsetting the southward flow of cooler NADW. Much of the heat lost to the atmosphere in the North Atlantic is derived from this cross equatorial heat transfer.

There is evidence that global warming has upset the normal flow of ocean currents in the South Atlantic. Some warmer South Atlantic waters are apparently flowing southward toward Antarctica and will eventually undermine and melt the vast ice shelves that are found extending off of the continent. As these ice shelves are consumed, land glaciers that feed them move seaward to replace the ice shelves and sea level is raised.

## 1.17   Ocean Acidification

Ocean acidification is defined as a reduction in pH. The oceans are not turning into roiling bodies of acid, although that would be the ultimate effect of ocean acidification. The World Ocean is undergoing a reduction of pH, therefore it is being acidified. Between 1751 and 2013 surface ocean pH is estimated to have decreased from approximately 8.25–8.08 (PNAS 2014). It is expected to drop by a further 0.3–0.5 pH units. A pH of 7 is neutral and below 7 is acidic. The current acidification is on a path to reach levels not seen in the last 65 million years, The ocean is still basic, chemically, but the effects of the reduction in pH are causing difficulty for some marine organisms like clams, mussels, oysters, corals, and others that secrete shell-forming calcium carbonate ($CaCO_3$) that they need for protection or body support.

With a lowered pH, these organisms are not able to extract calcium carbonate ($CaCO_3$) from sea water to build their shells and those that do have weaker shells. Ocean acidification is already being felt in oyster beds in Washington state, Oregon, and other areas around the world. Ocean acidification can only get worse with time as long as $CO_2$ is being taken up by the ocean. Ocean acidification is directly related to $CO_2$ emissions.

The following chemical reactions take place between ocean waters and $CO_2$ that result in acidification:

$$CO_2 + H_2O \leftrightarrow H_2CO_3 (\text{carbonic acid}) \qquad (1.1)$$

$$CO_2 + H_2O + CO_3^{2-} \leftrightarrow 2HCO^{3-} \qquad (1.2)$$

The chemical reactions that occur reduce seawater pH, carbonate ion concentration, and saturation states of biologically important calcium carbonate minerals (e.g., calcite and aragonite). Currently, the World Ocean is a gigantic sink for carbon dioxide.

Figure 1.18 shows the relationship between carbon dioxide increase, ocean pH decrease, and the increase in ocean $pCO_2$.

Carbon dioxide is soluble in cool ocean water. As ocean waters continue to warm, carbon dioxide will no longer be taken up by the ocean but ocean waters will become a source; ocean waters will then be a source for carbon dioxide, will accelerate global warming and thus cause numerous tipping points to be reached.

Increasing acidity is thought to have a range of harmful consequences, such as depressing metabolic rate and immune response in some organisms, and causing coral bleaching. Coral bleaching has received much attention from scientists because of its obvious deleterious effects on fragile reef ecosystems. Bleached corals are dead

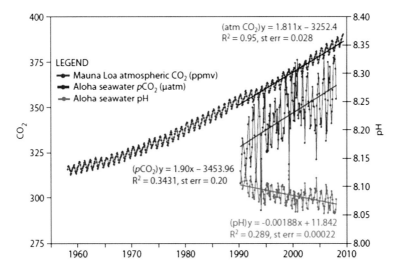

**Fig. 1.18** Annual variations in atmospheric $CO_2$, oceanic $CO_2$, and ocean surface pH. Strong trend lines for rising $CO_2$ and falling pH [from Feely et al. (2009)]

corals and bleached corals occur at the reef's surface. A coral reef grows upward on its dead relatives to remain in the photic zone because of symbiotic algae that coral can't live without. Most corals obtain the majority of their energy and nutrients from photosynthetic unicellular algae that live within the coral's tissue called zooxanthella (also known as Symbiodinium). Such corals require sunlight and grow in clear, clean, shallow water, typically at depths shallower than 60 m (200 ft).

There have been other times in Earth history that the ocean has been acidified. The most notable example is the Paleocene-Eocene Thermal Maximum (PETM), which occurred approximately 56 million years ago. For reasons that are currently uncertain, massive amounts of carbon entered the ocean and atmosphere, and led to the dissolution of carbonate sediments in all ocean basins. One possible explanation for PETM warming and the ocean acidification that occurred then was a sudden release of methane from methane hydrates on the ocean floor, in particular on the relatively shallow continental shelf areas.

Current rates of ocean acidification have been compared with the greenhouse event that occurred during PETM when surface ocean temperatures rose by 5–6 °C. No catastrophe was seen in surface ecosystems but bottom-dwelling organisms in the deep ocean experienced a major extinction. The current acidification is on a path to reach levels higher than any seen in the last 65 million years and the rate of increase is about ten times the rate that preceded the Paleocene–Eocene mass extinction. The current and projected acidification has been described as an unprecedented geological event.

A study entitled "Inhospitable Oceans," published in the peer-review journal *Nature Climate Change* on August 26, 2013, was based on examinations of five key components of ocean ecosystems: corals, echinoderms, mollusks, crustaceans and fish. All were found to be adversely affected by acidification: crustaceans were more resilient, while corals, mollusks and echinoderms were worst affected. The direct effects on fish were less clear.

## 1.18  Land-Atmosphere Interface

The land-atmosphere interface is, of course, where the atmosphere is in contact with the solid or glaciated Earth. This interface varies greatly due to the topography, nature of the materials making up the solid or glacial part of the interface, and the altitude. The land-atmosphere interface as a whole in the short term does not influence climate as does the ocean-atmosphere interface because of the lack of exchange of gases. However, the land-atmosphere interface is extremely important because of its effects on Earth's reflectivity (albedo). As the Sun's radiation strikes the Earth, an amount of this radiation is reflected back toward space. The percentage of light reflected back is the surface's albedo. Clean snow, sea ice, or newly formed glacial ice are nearly 100 % reflective and are said to have an albedo of 1.0 or just less than 1.0. A portion of Earth's surface that reflects 50 % of Sunlight that strikes it has an albedo of 0.5, and so on. Table 1.3 shows the albedo for some common Earth materials.

**Table 1.3**  Albedo of common Earth substances

| Material | Albedo |
|---|---|
| Fresh snow and ice | 0.7–0.9 |
| New concrete | 0.55 |
| Fresh asphalt | 0.04 |
| Green grass | 0.25 |
| Bare soil | 0.17 |
| Deciduous trees | 0.15–0.18 |
| Desert sand | 0.4 |

Albedo affects climate as the reflection of light from a surface has either a cooling or a warming effect. For example, in a region covered by clean glacial ice the effect is cooling because most of the solar radiation is reflected back toward space. As the glacial ice melts and the glacier recedes, or the glacial ice becomes dirty (with rock and soil materials or black carbon from industry and forest fires), the effect on the region is warming because the underlying land surface (rock or soil) has less reflectivity than the glacial ice and absorbs more Sunlight. The dirty ice is darker in color than clean ice so more heat is absorbed by the dirty ice making it melt more rapidly.

## 1.19  Paleoclimates and Paleoclimatology

Paleoclimates are those climates that existed on Earth in the past that most often were different from those that exist today. Lines of evidence of past climatic conditions are from the written word, such as the 'Little Ice Age' or the 'Medieval Warm Period' or from evidence that scientists call proxies (substitutes). Proxies are used as a substitute for hard data (instrumental data) where instrumental data do not exist.

Proxies that are used to reconstruct past climates are ice and sediment cores, the shells of microscopic organisms (mainly foraminifera), cave deposits, corals, tree rings, varves (annual layers of sediment), and certain sedimentary rock types. Proxies have told climate scientists a great deal about ancient climates, especially about the past 2.58 million years. The study of ancient climates is paleoclimatology. Figure 1.19 shows results of an ice core proxy for the past 650,000 years for $CO_2$. The plot shows the atmospheric concentration of $CO_2$ in the ice-core air bubbles that occupy pockets preserved within the ice.

## 1.20  Rising Sea-Level

Water expands as it warms so rising sea-level (eustatic or world-wide) is in part due to ocean water expanding. Fresh water is being added to the ocean from glacial ice that is melting and finding its way to the sea. The U.S. Geological Survey (USGS) has estimated that if all the ice on Earth melted, sea-level world-wide

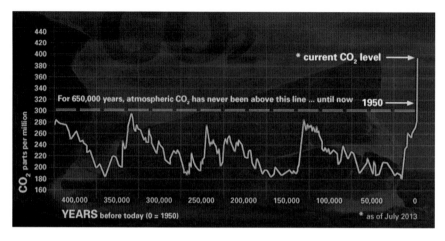

**Fig. 1.19** This plots atmospheric $CO_2$ concentration synthesizing ice core proxy data 650,000 years in the past capped by modern direct measurements (NASA, Public Domain)

would rise 260 ft, displacing millions of people that now live at or near sea-level. Such an event would involve the total collapse of the Greenland and Antarctic ice caps plus the melting of all other glaciers throughout mountainous areas of the world. This is not likely to happen suddenly as some glacial ice in Antarctica is at least two miles thick and in some places near the center of the continent, the ice thickness has been measured to be over 15,600 ft thick.

During the period 1901–2010, the rate of global averaged sea level rise was 0.07 in. (0.1753 mm) per year, which accelerated to 0.13 in. (3.25 mm) per year between 1993 and 2010. The rise in sea level, although modest averaged globally, is sufficient to cause severe storm surges in many parts of the world. According to the IPCC AR5 2013, the mean rate of sea level increase was 1.7 mm (0.068 in.) per year with a very likely range between 1.5 mm (0.06 in.) to 1.9 mm (0.076 in.) between 1901 and 2010 and this rate increased to 3.2 (0.128 in.) with a likely range of 2.8 (0.112 in.) to 3.6 mm (0.144 in.) year between 1993 and 2010.

The rate of sea level rise varies from place to place depending on local geological conditions. At an active tectonic plate boundary, such as along the California coastline of the U.S., sudden Earth movements may displace shorelines vertically tens of feet.

In the eastern part of the U.S. much of the shoreline south of Long Island, NY is sinking. North of Long Island the shoreline is rising. Both of these movements are related to the Last Glacial Maximum (LGM) when a massive sheet of ice, estimated to have been as much as 2 miles thick, extended south to Long Island, NY. This sheet of ice caused the crust to bow upward south of Long Island, analogous to weight being placed on a mattress. The crust north of Long Island was under the tremendous weight of the ice and was depressed. Now the crust is adjusting to the loss of the weight of the ice by moving vertically down to the south and up to the north of Long Island. Long Island, NY represents the terminal moraine of the last glacial advance and marks the southern end of the ice sheet. After reaching the latitude of Long Island, NY the massive continental glacier receded to the north.

## 1.21  The Cryosphere and Melting Glaciers

The cryosphere consists of all ice formed in nature including the ice caps of Greenland and Antarctica, all mountain or alpine glaciers in mountainous areas of the world, permafrost, sea ice, icebergs, and ice that forms on rivers, lakes, and ponds in the winter in middle and high latitudes. Much of this ice is melting.

Glaciers are bodies of ice that are found today in high latitudes (near the geographic poles) and at high altitudes. Most of this glacial ice is in the process of wasting away and adding melt-water to the ocean. As glacial ice continues to melt, sea-level will continue to rise.

Permafrost is frozen ground (for at least two consecutive years) and is found in high latitudes, high elevations, and around extant glaciers. A good amount of permafrost is thought to be left over from the last 'ice age.'

Permafrost is melting, especially rapidly at its lower latitudes and elevations, but this is not happening overnight. Changes in temperature at the surface take time to impact permafrost at depth; according to the Geological Survey of Canada (GSC), "for thick permafrost this lag may be on the order of hundreds to thousands of years; for thin permafrost, years to decades" (Utting et al. 2007). The distribution of permafrost worldwide is seen in Fig. 1.20. Approximately 1,670 petagrams (Pg; see Appendix B) of soil carbon are estimated to be stored in soils and permafrost of high latitude ecosystems.

Russia has a long-term permafrost monitoring system. There has been a push to extend current monitoring programs and enlarge their scope. Programs such as the Global Terrestrial Network for Permafrost (GTNP) are working to organize data collection so that there is a global network for detecting and monitoring changes in permafrost regions, and predicting climate change's impact on these affected

**Fig. 1.20**  Permafrost distribution in the Arctic (*image credit* Philippe Rekacewicz, 2005, UNEP/ GRID-Arendal Maps and Graphics Library based on International Permafrost Association (1998) Circumpolar Active-Layer Permafrost System (CAPS), version 1.0)

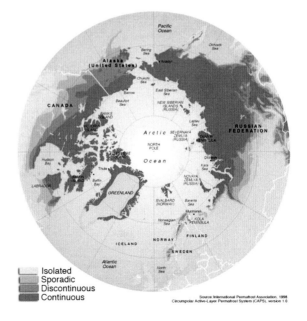

areas. Advances in spatial analysis have contributed greatly to depicting areas and depths of permafrost.

The University of Alaska–Fairbanks has an active permafrost laboratory. Dr. Vladimir Romanovsky, Professor of Geophysics at the university, is a leading researcher on permafrost. There are others doing research on permafrost and the potential release of methane as the permafrost continues to melt.

### 1.21.1 Glaciers

The cryosphere is the largest reservoir of fresh water on Earth. It includes all the ice formed in nature. It also provides a direct indicator of the behavior of sea level. As glaciers melt, sea level rises. If glaciers expand, sea level is lowered.

Snow is precipitation in the form of ice crystals. It originates in clouds when temperatures are below the freezing point and water vapor in the atmosphere condenses directly into ice without going through the liquid stage.

Once an ice crystal has formed, it absorbs and freezes additional water vapor from the surrounding air, growing into a snow crystal which then falls to Earth. Snow is not frozen rain but is frozen water vapor.

Snow is important for climate because it affects the amount of Sunlight reflected back to space (Earth's albedo). Reflected Sunlight causes the Earth's climate to cool. When snow or ice melts, darker land and ocean absorb more of the Sun's rays and Earth's climate warms. Land covered in fresh snow reflects nearly 100 % of Sunlight (an albedo close to 1.0). Snow that has settled and become granular is called firn. It is firn that turns into glacial ice.

Melting glaciers are a great concern for millions of humans that live in areas where their main source of water for drinking, bathing, and crop irrigation is glacial meltwater. This is the case for many of those who live in mountainous regions that still contain glaciers, such as the Alps, Himalayan, Andes Mountains and Rocky Mountains.

Glaciers in mountainous regions occupy former stream valleys and are referred to by any one of several names, such as mountain, alpine, and valley glaciers. These glaciers are rivers of ice that flow downslope under the influence of gravity. Glaciers are constantly on the move, even when their terminus is stationary or moving upslope (receding). The glacial ice is constantly moving toward the terminus and carrying rock and soil with it. A common analogy is that a stable glacier, whose terminus is not moving, moves material as does a conveyor belt.

### 1.21.2 Greenland Ice Sheet

The Greenland Ice Sheet covers approximately 1,710,000 km² (660,000 square miles). It is the second largest ice sheet on Earth, exceeded in size by only the Antarctic ice sheet. It currently covers about 80 % of the surface of Greenland (Fig. 1.12).

The thickness is generally more than 2 km (1 mile) and over 3 km (1.9 miles) at its thickest point. It is not the only ice mass of Greenland; isolated glaciers and small ice caps cover between 76,000 and 100,000 km$^2$ (29,000 and 39,000 square miles) around the periphery. Many climate scientists think that climate warming is at, or has passed, a "tipping point" where the entire ice sheet will melt and there is nothing humans can do about it. If the entire 2,850,000 cubic kilometers (684,000 km$^3$) of ice were to melt, it would lead to a global sea level rise of approximately 7.2 m (24 ft).

For several days in July 2013, Greenland's surface ice cover melted over a larger area than at any time in more than 30 years of satellite observations. Nearly the entire ice cover of Greenland, from its thin, low-lying coastal edges to its two-mile-thick center, experienced some degree of melting at its surface, according to measurements from three independent satellites analyzed by NASA and university scientists.

The Greenland Ice Sheet has experienced record melting in recent years and will contribute substantially to sea level rise as well as to possible changes in ocean circulation in the future. Changes to ocean circulation will change weather patterns throughout the world.

The Greenland Ice Sheet formed in the middle Miocene (see Appendix A) by coalescence of ice caps and glaciers. There was an intensification of glaciation during the Late Pliocene which continued into the Pleistocene.

## 1.22 More Violent Storms

Individual storms (cyclones, hurricanes, tornadoes, precipitation events) are increasing in intensity due mainly to global warming. Hurricanes Sandy and Katrina are two of the latest to impact North America and Typhoon Haiyan (2013) in the Philippines are examples of cyclones due to a warming Earth.

Britain braced for further travel and power network disruption as an Atlantic storm battered the British coast the weekend of February 8, 2014. Ruth, the worst of a "conveyor belt of storms" brought winds of up to 80 mph and as much as 40 mm (1.6 in.) of rain.

The "absolute monster" storm means waves of up to 35 ft smashing the Cornish coast, forecasters from surf website magicseaweed.com said.

The south of England and Wales was covered by a Met office severe weather warning for Saturday (February 8, 2013) for gales, torrential rain and huge waves. Although it is difficult to assign the cause of individual storms as being the result of global warming, high ocean temperatures, increased rates of evaporation leading to more moisture in the atmosphere, have increased the severity of recent storms.

The increase in Earth's temperature by as much as 1 °C since the start of the Industrial Revolution has caused new weather patterns to develop on the planet. Weather is being influenced greatly by global warming and meteorologists, climatologists, and other scientists are suggesting that Earth is in a 'new-normal' period of continuing weather patterns. Loss of sea ice in the Arctic has caused the Jet Stream to develop very large bends or meanders as it flows over the Northern Hemisphere causing long spells of extreme temperatures, precipitation, and drought especially over large areas of North America.

## 1.23 Deforestation

Deforestation is another of humankind's activities that contribute to global warming. Forests are carbon sinks. They soak up large quantities of carbon dioxide. Cutting down the trees of a forest, burning them, or leaving them to decay releases carbon dioxide to the atmosphere.

It's difficult to convince the residents of the Amazon basin, Indonesia, and other tropical regions of the world to stop cutting down trees when the forests are worth more dead than alive. Conservation costs money, while profits from timber, charcoal, pasture and cropland drive people to cut down forests. Exacerbating global warming isn't the only negative impact of tropical deforestation. It also wipes out biodiversity: More than half of the world's plant and animal species live in tropical rainforests.

Deforestation most likely first became common when humans learned to use fire as a tool to drive animals from their hiding places within forests so that the animals could be captured and eaten. Deforestation increased during the Agricultural Revolution as vast woodland areas of the world were converted to farmland. This practice is still occurring, especially in the rainforests of Africa, Indonesia, and the Amazon region of South America.

The United Nations Framework Convention on Climate Change (UNFCCC) has determined that the overwhelming direct cause of today's deforestation is agriculture (subsistence farming is responsible for 48 %; commercial agriculture is responsible for 32 %; logging is responsible for 14 %; and fuel wood removals make up the remaining 6 %). Reports from deforestation areas tell us that vast areas of the forest are being cut down to make way for palm oil and other plantations.

Clear-cutting, logging, and cutting down trees in general are active contributors to warming of the planet. Trees and other vegetation are sinks for carbon dioxide as they remove it from the atmosphere and use it in photosynthesis. Deforestation has contributed to global warming by stopping trees from growing and taking carbon dioxide from the atmosphere for photosynthesis. Decaying vegetation and forest fires release carbon dioxide back into the atmosphere.

Forest soils are moist, but without protection from the Sun by tree cover they quickly dry out. Trees also help perpetuate the water cycle by returning water vapor back into the atmosphere. Without trees to fill these roles, many former forest lands can become barren deserts.

## 1.24 Desertification

Global warming has begun to upset well-established weather patterns throughout the world, especially in the Northern Hemisphere. Rainfall patterns are one example. Areas that have been deserts may be receiving more rainfall and wetter areas may be receiving less rainfall. Also, the reverse may be occurring in some areas, i.e., desert areas may be getting less rainfall and wetter areas may be getting more precipitation. It is also true that some areas of desert are expanding.

Desertification results in soil degradation, the removal of soil by erosion and wind causing the area or region to change from arable to non-arable; crops can no longer be grown in such soils. This is a serious problem for subsistence farmers who live near areas of desertification if the desert is expanding in area. As desertification expands in aerial extent, these subsistence farmers will no longer be able to grow their own food and will be forced to move or change vocations.

Although desertification may include the encroachment of sand dunes on land, it doesn't necessarily refer to the advance of deserts. It refers to the persistent degradation of dry-land ecosystems by human activities and may include depleting soil nutrients by unsustainable farming, mining, overgrazing, clear-cutting of land, and by climate change.

## 1.25  Species Migration

Because of global warming, species are migrating to higher elevations and higher latitudes. Birds and insects are staying in areas for longer periods of time, mainly during fall and winter seasons due to rising temperatures. They are also arriving earlier in the spring due to earlier springs in temperate latitudes. Rising temperatures are also causing disruptions in natural cycles such as breeding seasons for animals and budding season for plants.

From chipmunks to spiders, animals are striving to cope with the effects of warmer temperatures and plants are not immune. Tropical Andean tree species are shifting roughly 8–12 vertical feet (2.5–3.5 m) a year on average, the arboreal equivalent of a 100-m dash. Yet for those trees to remain in equilibrium with their preferred temperatures, they need to migrate more than 20 vertical feet a year, so say the plant ecologists and botanists.

DNA evidence suggests the European wasp spider is evolving into a new form and is moving to cooler regions to set up home in parts of northern Europe, while chipmunks living in Yosemite National Park in California are moving to higher, cooler altitudes.

The changes are due to the effects of global warming and as habitats move, populations of animals that have previously not crossed paths, are mixing and have the potential to spread and adapt in new ways. These animals are those that can migrate and not all animals can.

What happens when these animals and plants can go no higher in altitude or latitude?

## 1.26  Species Extinctions

The Earth has experienced five major episodes of mass extinctions during its history. We are perhaps witnessing the beginning of the sixth mass extinction. "The Sixth Extinction," a new book by Elizabeth Kolbert was published by the time of this writing (February 2014).

Species are becoming extinct at the present time most likely more rapidly than has happened during the history of humankind. It is impossible to know the rate of extinction and number of species going extinct today because biologists know that all species on Earth have not yet been identified, described, and named.

## 1.27 Changing Seasons and Disruption of Life Cycles

As the planet continues to get warmer, seasons of the year change. Spring and summer last longer and fall and winter become shorter. The life cycle of organisms is also changing. For example, some species are able to survive winters that they could not do prior to warming. Others may mature before their food supply matures.

Throughout the Rocky Mountains in the western U.S. and Canada, a pine bark beetle is destroying trees by the millions. Although bark beetle infestations are a regular force of natural change in forested ecosystems, several of the current outbreaks, occurring simultaneously across western North America, are the largest and most severe in recorded history. Some beetles have shifted to completing their development in a single year rather than 2 or even 3 years. Assuming other inputs to the system remain constant, this decrease in generation time translates to a doubling or tripling in the rate of population growth.

Many organisms' life cycles are regulated by climate; temperature, drought, precipitation, to name a few consequences of climate change. Shelford's Law of Tolerance is a principle developed by American zoologist Victor Ernest Shelford in 1911. It states that an organism's success is based on a complex set of conditions and that each individual or population has a certain minimum, maximum, and optimum environmental factor or combination of factors that determine success. When climate changes, minimum or maximum tolerances for many organisms are exceeded and the organisms are likely to migrate or to become extinct.

## References and Additional Reading

Abrupt Impacts of Climate Change: Anticipating Surprises (2013) National Research Council, National Academies of Science. ISBN 978-0-309-28773-9
Balmaseda M et al (2013) Evaluation of the ECMWF ocean reanalysis system ORAS4. Q J R Meteorol Soc 139:1132–1161. http://dx.doi.org/10.1002/qj.2063
Buckley LB, Roughgarden J (2004) Biodiversity conservation: effects of changes in climate and land use. Nature 430(6995). doi:10.1038/nature02717
CDIAC (2012) Carbon Dioxide Information Analysis Center (CDIAC). CDIAC, Oak Ridge
Climate change: evidence and causes. Proceedings of the National Academy of Sciences and the UK Royal Society
Collins WD et al (2006) Radiative forcing by well-mixed greenhouse gases: estimates from climate models in the intergovernmental panel on climate change (IPCC) fourth assessment report (AR4). J Geophys Res 111(D14317): D14317. Bibcode: 2006JGRD.11114317C. doi:10.1029/2005JD006713

Dahl-Jensen D et al (2013) Eemian interglacial reconstructed from a Greenland folded ice core. Nature 493(7433):489–494

Farmer GT, Cook J (2013) Climate change science: a modern synthesis. Springer, Dordrecht and New York

Feely R, Doney S, Cooley S (2009) Present conditions and future changes in a high-$CO_2$ world. Oceanography 22:36–47

Fröhlich C, Lean J (1998) The Sun's total irradiance: cycles and trends in the past two decades and associated climate change uncertainties. Geophys Res Lett 25:4377–4380

Gazeau F et al (2013) Impacts of ocean acidification on marine shelled molluscs. Marine Biol Zitiert durch: 9 - Ähnliche Artikel - Alle 6 Versionen

Hansen J et al (2011) Earth's energy imbalance and implications. Atmos Chem Phys 11:13421–13449. doi:10.5194/acp-11-13421-2011

Hansen J et al (2013a) Climate sensitivity, sea level and atmospheric carbon dioxide, vol 371. Royal Society Publishing. doi:10.1098/rsta.2012.0294

Hansen J et al (2013b) Climate forcing growth rates: doubling down on our Faustian bargain. Environ Res Lett 8:011006. doi:10.1088/1748-9326/8/1/011006

Hansen J et al (2013c) Assessing 'Dangerous Climate Change': required reduction of carbon emissions to protect young people. Future generations and nature. PLOS ONE 8:e81468

Harte J et al (2004) Biodiversity conservation: climate change and extinction risk. Nature 430(6995). doi:10.1038/nature02718

Introduction to carbon capture and storage—carbon storage and ocean acidification activity. Commonwealth Scientific and Industrial Research Organisation (CSIRO) and the Global CCS Institute

IPCC AR4 20047 (2007) Chapter 8: climate models and their evaluation. The IPCC working group I fourth assessment report (2007)

IPCC AR5 report (September 30, 2013) The physical science basis

IPCC special report carbon dioxide capture and storage summary for policymakers. Intergovernmental Panel on Climate Change

Keeling CD, Whorf TP (2004) Atmospheric $CO_2$ from continuous air samples at Mauna Loa Observatory. Carbon Dioxide Information Analysis Center, Oak Ridge National Laboratory, Hawaii

Kiehl JT, Trenberth KE (1997) Earth's annual global mean energy budget. Bull Amer Meteor Soc 78:197–208

Kosaka Y, Xie S-P (2013) Recent global-warming hiatus tied to equatorial Pacific surface cooling. Nature (published online 28 Aug 2013). doi:10.1038/nature12534

Marcott SA et al (2013) A reconstruction of regional and global temperature for the past 11,300 years. Science 339(6124):1198–1201. doi:10.1126/science.1228026

Mason J (2013) A rough guide to the components of Earth's Climate System. www.skepticalscience.com. Posted 7 Oct 2013

Mora C (2013) The projected timing of climate departure from recent variability. Nature 502:183–187. doi:10.1038/nature12540

Moritz C (2008) Impact of a century of climate change on small-mammal communities in Yosemite National Park, USA. Science 322:261–264

National Aeronautics and Space Administration (NASA) (2014) What is the difference between weather and climate? http://www.nasa.gov/mission_pages/noaa-n/climate/climate_weather.html

National Oceanic and Atmospheric Administration (NOAA)—Earth System Research Laboratory (ESRL) (2013) Trends in carbon dioxide

Skinner BJ, Porter SC (2004) Dynamic earth: an introduction to physical geology—with CD (5TH 04), paperback. Wiley, New York. ISBN13: 978-0471152286 ISBN10: 0471152285

Solomon S, Qin D, Manning M, Chen Z, Marquis M, Averyt KB, Tignor M, Miller HL (eds) (2007a) Intergovernmental panel on climate change (IPCC), climate change 2007: the physical science basis. Cambridge University Press, Cambridge, 996 pp

Solomon S, Qin D, Manning M, Chen Z, Marquis M, Averyt KB, Tignor M, Miller HL (eds), IPCC AR4 WG1 (2007b) Climate change 2007: the physical science basis, contribution of working group I to the fourth assessment report of the intergovernmental panel on climate change. Cambridge University Press. ISBN 978-0-521-88009-1 (pb: 978-0-521-70596-7)

Trenberth KE, Fasullo J, Kiehl J (2009) Earth's global energy budget. Bull Am Meteor Soc 90:311–324

Trenberth KE, Fasullo JT (2013) Changes in the flow of energy through the Earth's climate system. Meteorol Z 18(4):369–377

University of Toronto (2013) New long-lived greenhouse gas discovered: highest global-warming impact of any compound to date. Science Daily (9 Dec 2013). www.sciencedaily.com/release s/2013/12/131209124101.htm

Utting DJ, Gosse JC, Hodgson DA, Trommelen MS, Vickers KJ, Kelley SE, Ward B (2007) Geological Survey of Canada, Natural Resources Canada

Van Vuuren DP et al (2011) The representative concentration pathways: an overview. Clim Change 109:5–31. doi:10.1007/s10584-011-0148-z

Wayne GP (2013) Beginner's guide to representative concentration pathways. Available online at the following website: www.skepticalscience.com

Zachos JC et al (2005) Rapid acidification of the ocean during the Paleocene-Eocene thermal maximum. Science 308(5728):1611–1615. doi:10.1126/science.1109004 (PMID 15947184)

# Chapter 2
# Status of Climate Change Research

**Abstract** This chapter contains an overview of research in climate change science. Research ongoing and recent publications concerning each topic in Chap. 1. Topics of climate change research, such as temperature change on land and sea, impacts of climate change on glacial ice, the extent of permafrost and its carbon content, research into the carbon cycle, ocean acidification and circulation, and General Circulation Climate Models (GCMs) as well as the biology affected by climate change are discussed with references given.

**Keywords** CMIP5 · Permafrost · Moraine · PDO · Weart · Fourier · NCDC · BEST · LGM · Holocene · Anthropocene · Atmosphere · Stratosphere · Sangamonian · Eemian · www.skepticalscience.com · John Cook · Petagram · CERES · ICOS · 0.9 °C · ICOADS · David Archer · Fast carbon · Slow carbon · Carbon · Carbon cycle · TERRA · Albedo · Argo · MODIS · GHCN · *Homo sapiens* · ENSO · MSUs · Microwave sounding units · Solubility pumps · Biological · Solubility · Cryosphere · Scripps · La Jolla · GTNP · Tibetan plateau · NASA · NOAA · NSF · Virkisjökull · Greenland · Antarctica · Sea ice · Larsen B · PETM · Pleistocene · Pliocene · Anthropocene · GRACE · La Niña · $^{12}C$ · $^{13}C$ · $^{14}C$ · CMIP5 · Atlantic · Pacific · Indian · GtC · Tyndall · Arrhenius

## 2.1 Introduction

Climate change research is healthy world-wide and there is a lot of it being conducted, only a small portion of which can be covered here. Climate research is not new. It has been conducted since Jean Baptiste Joseph Fourier did groundbreaking research on heat transfer and realized that the Earth would be much colder without the atmosphere. Earth's temperature usually given without the atmosphere is $-15$ to $-18$ °C. Other scientists followed Fourier in researching various aspects of climate and the Earth's temperature. John Tyndall conducted research that proved that carbon dioxide and water vapor were opaque to heat radiation. Svante Arrhenius,

© The Author(s) 2015
G.T. Farmer, *Modern Climate Change Science*,
SpringerBriefs in Environmental Science, DOI 10.1007/978-3-319-09222-5_2

a Swedish chemist, calculated how much the temperature would rise if $CO_2$ in the atmosphere doubled.

Svante Arrhenius was the first person to predict that emissions of carbon dioxide from the burning of fossil fuels and other combustion processes were large enough to cause global warming. His work is currently seen as the first demonstration that global warming should be taken as a serious possibility. Because he lived in Sweden, he thought that warming would be a good thing.

## 2.2  Research in Weather and Climate

The differences between weather and climate are treated in Sect. 1.1. Studies of climate show long-term trends of 30 years or more. Climate models treat the atmosphere, geosphere, biosphere, hydrosphere, and cryosphere as totally integrated aspects of climate. Weather models predict weather conditions tomorrow and possibly a few weeks in advance. Climate models are global and regional. Weather models are regional and local. Therefore, globally rising temperatures since 1750 or 1880 to the present are in the domain of climate, not weather. Storms, hurricanes, and the day-to-day temperatures are weather.

Global average temperatures for 2013 have recently been published by the BEST study in Berkeley, California USA, NASA/GISS, NOAA/NCDC, the U.K.'s Met Office and the University of East Anglia Climate Research Unit's HadCRUT4, and Cowtan and Way's recent publication in the Quarterly Journal of the Royal Meteorological Society. An analysis of these data is given by Stefan Rahmstorf at www.realclimate.org and is not repeated here. Suffice it to say that 2010 and 2005 remain the warmest years since records began in the 19th Century. 1998 ranks third in two records, and in Cowtan and Way's 2014 analysis, which interpolates the data-poor regions in the Arctic, Africa, and Antarctica, 2013 is warmer than 1998. 1998 was a very significant El Niño year, which generally warms the climate, and 2013 is neutral.

Tropospheric and lower stratospheric temperature data are collected by NOAA's TIROS-N polar-orbiting satellites and adjusted for time-dependent biases by the Global Hydrology and Climate Center at the University of Alabama in Huntsville (UAH). An independent analysis is also performed by Remote Sensing Systems (RSS) and a third analysis has been performed by Fu and Johanson (2005) of the University of Washington (UW).

Since 1978 Microwave Sounding Units (MSUs) measure radiation emitted by the Earth's atmosphere from NOAA polar orbiting satellites. The different channels of the MSU measure different frequencies of radiation proportional to the temperature of broad vertical layers of the atmosphere. The analysis of the satellite temperature record begins in 1979.

Fu and Johanson (2005) developed a method for quantifying the stratospheric contribution to the satellite record of tropospheric temperatures and applied an adjustment to the UAH and RSS temperature record that attempts to remove the satellite contribution (cooling influence) from the middle troposphere record. This method results in trends that are larger than those from the satellite source.

A comprehensive inter-comparison (Seidel et al. 2004) showed that five radiosonde data sets yielded consistent signals for higher-frequency events such as ENSO, the Quasi-Biennial Oscillation (QBO) and volcanic eruptions, but inconsistent signals for long-term trends.

Radiosonde data must be adjusted for comparison to satellite temperature data. Analysis of radiosonde data can be found in NCDC's monthly Upper Air State of the Climate Report and the Bulletin of American Meteorology Society's Annual State of the Climate Report.

Cowtan and Way 2014 reported on an independent analysis of temperature data using the HadCRUT4 data set. HadCRUT4 is widely quoted as a measure of global warming. However observations in the dataset are only available for about 84 % (five sixths) of the planet. The omitted region includes the Arctic, which is warming at two and a half times faster than U.K.'s Met Office estimates over the past 16 years. As a result, HadCRUT4 underestimates the rate of warming in recent years. The Cowtan and Way dataset uses satellite data to fill in the gaps in the Met Office data (HadCRUT4).

## 2.3  The Intergovernmental Panel on Climate Change

The Intergovernmental Panel on Climate Change (IPCC) was introduced in Sect. 1.2. It is part of the United Nations and consists of a body of scientists with a record of contributions to climate science. Their task is to review and assess all the research that has been published in the time since their previous report. They do no original research and they do no monitoring for their reports. The reports are published about every 6 or 7 years. However, their reports form the basis for and a reference for many lines of climate change research.

## 2.4  Representative Concentration Pathways

The IPCC, in its AR5 2013 report introduced the use of Representative Concentration Pathways (RCPs), each of which represents an emission scenario; a new set of scenarios that replaces the Special Report on Emissions Scenarios (SRES) standards employed in two of the previous IPCC Assessment Reports (TAR and AR4). Emission scenarios are used primarily for projections by varying input that satisfies the conditions as set forth by the RCPs. The RCPs are especially important to the climate modeling groups working with large computer models as they provide a baseline of criteria that can be used by multiple groups. Standardizing the scenarios means that each group starts with the same criteria.

## 2.5  Radiative Forcing

Radiative forcing (RF) was defined in Sect. 1.4. In Fig. 1.3 we saw that carbon dioxide was the most powerful anthropogenic forcing agent.

Radiative forcing of the climate system involves every aspect of the system from space to the Earth's interior (e.g., volcanoes). To cite research ongoing in radiative forcing would involve a separate volume. As a result, a decision has been made to emphasize some of the research being done at the U.S. National Aeronautical and Space Administration's Goddard Institute of Space Studies (NASA/GISS).

### 2.5.1  Research in Radiative Forcing

Scientists at NASA/GISS have the Atmospheric Chemistry and Climate Model Intercomparison Project (ACCMIP) that evaluates the role of atmospheric chemistry, both gases and aerosols, in driving climate change radiative forcings. In particular, the intercomparison is designed to facilitate analyses of the driving forces of climate change in simulations being performed in the Climate Model Intercomparison Project phase 5 (CMIP5) in support of the IPCC AR5 and beyond.

The Goddard Institute Surface Temperature (GISTEMP) Analysis group considers surface air temperature change as a primary measure of global climate change in response to one or more forcings. The GISTEMP project started in the late 1970s to provide an estimate of the changing global surface air temperature which could be compared with the estimates obtained from climate models simulating the effect of changes in atmospheric carbon dioxide, volcanic aerosols, and solar irradiance. The continuing analysis updates global temperature change from the late 1800s to the present.

The Climate Impacts research group at NASA/GISS seeks to improve understanding of how climate affects human society.

Dr. James Hansen, who many refer to as the "Father of Modern Climate Science," was the Director of NASA/GISS until his retirement in April 2013. Hansen will remain an activist on climate change. He has published more groundbreaking papers in peer-reviewed journals and books and given more public presentations and interviews on climate change than anyone else to date. His publications appear in the "Additional Reading" lists more often than any other.

Dr. Gavin Schmidt, a well-known climate modeler and one of the founders of realclimate.org, is an employee of NASA/GISS. His main research interest lies in understanding the variability of the climate, both its internal variability and the response to external forcing. In particular, how changes related to varying forcings relate to variations due to unforced climate variability such as oscillations in the ocean's deep thermohaline circulation that affect ocean heat transports or atmospheric modes of variability like the North Atlantic Oscillation (NAO). He mainly uses large-scale GCMs for the atmosphere and ocean to investigate these questions.

There is also a Climate Modeling Group at NASA/GISS that develops and does research using GCMs.

## 2.6  Earth's Energy Imbalance and Energy Flow

Earth's "energy imbalance and energy flow" was introduced in Sect. 1.5. Figure 1.4 illustrates the energy flow through the climate system.

Earth has been in radiative imbalance, with more energy from the Sun entering than energy (heat) exiting the top of the atmosphere since at least the 1970s when satellite monitoring began. It is virtually certain that Earth has gained substantial energy from 1971 to 2010. The estimated increase in energy inventory between 1971 and 2010 is $274 \times 10^{21}$ J, with a heating rate of $213 \times 10^{12}$ W from a linear fit to the annual values over that time period.

Improvements in satellite data and interpretation has led to new analyses for energy, incoming and outgoing, at the top of the atmosphere (TOA).

### 2.6.1  Research into Earth's Energy Imbalance and Energy Flow

Since 2007, knowledge of the magnitude of radiative energy fluxes in the climate system has improved, requiring an update of the global annual mean energy balance. Energy exchanges between Sun, Earth and Space are observed from space-borne platforms such as the Clouds and the Earth's Radiant Energy System (CERES) and the Solar Radiation and Climate Experiment (SORCE) which began data collection in 2000 and 2003, respectively. The total solar irradiance (TSI) incident at the TOA is now much better known, with the SORCE Total Irradiance Monitor (TIM) instrument reporting uncertainties as low as 0.035 %, compared to 0.1 % for other TSI instruments. During the 2008 solar minimum, SORCE/TIM observed a solar irradiance of $1{,}360.8 \pm 0.5$ W/m$^2$ compared to $1{,}365.5 \pm 1.3$ W/m$^2$ for instruments launched prior to SORCE and still operating in 2008.

NASA/GISS does the bulk of the research on energy flow within the climate system. Other research is done using satellite data provided by NASA at the following institutions: University of Hawaii, the Pacific Marine Environmental Laboratory in Seattle, National Center for Atmosphere Research (NCAR), the University of Reading, U.K., and the University of Miami, U.S.

There are many other agencies and institutions using these data and they are too numerous to mention individually.

Historically, Jim Hansen has been one of the leading researchers of Earth's energy imbalance and has published widely about the imbalance and energy flow throughout the climate system. Kevin Trenberth, and J.T. Fasullo of NCAR are also leading researchers.

## 2.7  Earth's Rising Temperatures

The subject of global temperatures and the rising trend were introduced in Sect. 1.6.

The land surface of Earth has been warming along with the lower atmosphere since the Industrial Revolution and we have instrumental records since 1880. The land surface directly affects climate due to its variations of albedo (degree of reflectivity). For example, a land surface covered by relatively clean glacial ice (high albedo) will reflect more of the Sun's energy than a darker ocean (lower albedo). The darker ocean will absorb more energy and thus add more heat to the Earth's climate system. The albedo of common Earth materials is given in Table 1.3.

A rise in global average surface temperatures is the best-known indicator of a warming climate change. Although each year and even each decade is not always warmer than the last, global surface temperatures have warmed substantially since 1900. Warming land temperatures correspond closely with the observed warming trend over the ocean. Warming oceanic air temperatures, measured from aboard ships, and temperatures of the sea surface itself also coincide, as borne out by many independent analyses.

Land surface temperature is not the same as the air temperature that is included in the daily weather report. Land surface temperatures can vary throughout the year, but equatorial regions tend to remain consistently warm, and Antarctica and Greenland remain consistently cold. Altitude plays a clear role in temperatures, with mountain ranges like the Himalayas, Alps and North American Rockies colder than other areas at the same latitudes.

NASA obtains land surface temperature data from the Moderate Resolution Imaging Spectroradiometer (MODIS) on NASA's TERRA satellite (see Fig. 2.1 for a snapshot of data from TERRA/MODIS).

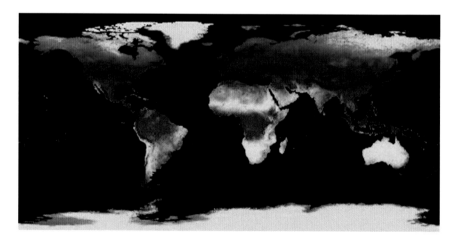

**Fig. 2.1** This map shows the temperature of Earth's lands during the daytime as mapped by NASA's TERRA/MODIS (from NASA, public domain)

Scientists can use a thermometer to measure the temperature of any single place and can measure the temperature of the whole world from space using instruments carried on satellites. However, records from satellites date only from 1978.

Scientists want to know the land's temperature for many important reasons. For example, in places where it is too hot or too cold, food crops may not live long enough to reach harvest. Land temperature also influences weather and climate patterns. Land surface temperature (LST) is the mean radiative "skin" temperature of an area of land resulting from the mean balance of solar heating and land-atmosphere cooling fluxes. LST is a basic determinant of the terrestrial thermal behavior, as it controls the effective radiating temperature of the Earth's surface. It determines the surface air temperature and the longwave radiation between the surface and the atmosphere, and is influenced by various surface-atmosphere boundary conditions, such as precipitation and albedo. In addition it influences the partitioning of energy into ground, sensible, and latent heat fluxes.

LST is more closely related to the physiological activities of leaves in vegetated areas, and to soil moisture in sparsely vegetated areas. Differences between simultaneous measurements of LST and air temperature can be as a much as 20 K (Kelvin), with LST exhibiting both a strong diurnal cycle and strong heterogeneity imposed by landscape variations.

There has been substantial land surface warming in the Arctic region during the past 50 or 60 years. Arctic land masses have been warming much faster than those of lower latitudes.

Research into temperatures trends, and computing the global temperature are discussed in the following section.

## 2.7.1  Current and Recent Research

Most research cites land temperatures as being measured. This is actually referring to the lower atmospheric temperature above land, and not the temperature of the land itself. The same is true of ocean temperatures. The readings over ocean are of the lower atmosphere over ocean water and not of the water itself. Measurements of the sea surface temperature are designated SST. The method of referring to lower atmospheric temperatures as "land" and "ocean" temperatures has become standard usage and is supposed to be intuitively obvious. However, this is not true to the uninitiated in climate science and should be clarified in publications aimed at a general audience.

Temperature data (with assessment) are the best indication of a warming planet. Earth's atmospheric temperature records from land stations come from just above the land surface mainly from "weather stations" scattered around the world. Figure 1.5 shows the temperature increase from 1880 to 2014 compared to the 1951 to 1980 global average.

Official temperature records are kept in the U.S. by the National Aeronautic and Space Administration/Goddard Institute of Space Sciences (NASA/GISS),

National Climatic Data Center (NCDC), the National Oceanic and Atmospheric Administration (NOAA), the Global Historical Climatology Network-Daily (GHCN-D), Global Historical Climatology Network-Monthly data set (GHCN-M), and in the U.K. by the Met Office Hadley Centre, and the University of East Anglia Climate Research Unit (CRU).

Other record-keeping agencies in countries throughout the world are too numerous to mention, but most of them get their data from the GHCN, except for the U.K. HadCRUT data and the BEST study (see below). A new dataset was recently compiled by Cowtan and Way (2014) using the Kriging method to interpolate HadCRUT temperature data from areas having few temperature reporting sites.

NASA's Goddard Institute of Space Sciences (GISS) is normally lumped together with NASA as NASA/GISS. NASA/GISS does an analysis of worldwide temperature data. The analysis uses satellite observed nightlights to identify measurement stations located in extreme darkness and adjust temperature trends of urban and peri-urban stations for non-climatic factors, verifying that urban effects on analyzed global change are small. They maintain a running record of any modifications made to the analysis which is available at their website: http://data.giss.nasa.gov/gistemp/.

The NASA/GISS GISTEMP analysis is reported by Dr. Reto Ruedy. Also participating in the GISTEMP analysis are Dr. Makiko Sato and Dr. Ken Lo. The research was previously led by Dr. James Hansen, now retired from NASA as of April 2013, but still active as an adjunct professor at Columbia University in New York City.

## 2.7.2 The BEST Study

The Berkeley Earth Surface Temperature (BEST) study is being conducted by a team at the University of California, Berkeley. The project was launched and chaired by Professor Richard Muller, a physicist at the University of California at Berkeley and a former "skeptic" concerning global warming. He was unsure of the methodology of determining global temperature and set out to see for himself whether or not it was accurate.

The BEST study addresses criticisms raised by other climate skeptics about how existing records of the Earth's average surface temperature have been compiled. The BEST group published their initial findings in 2012, which addressed concerns raised by legitimate skeptics about records of surface temperatures, including the "urban heat island effect" and poor weather station siting. These issues were found to not have a significant effect on the global land surface temperature record.

The latest research from the BEST team confirms that the Earth's average land temperature has risen by roughly 0.9 °C since the 1950s and by 1.5 °C over the last 250 years. This is in excess of other existing records, which put average global land temperature rise over the last 50 years at 0.81–0.93 °C.

The BEST study combined 1.6 billion temperature reports from 16 preexisting data archives. After eliminating duplicate records, the current archive contains over 39,000 unique stations. This is roughly five times the 7,280 stations found in the Global Historical Climatology Network Monthly data set (GHCN-M) that has served as the focus of many climate studies. The GHCN-M is limited by strict requirements. The BEST team developed new algorithms that reduce the need to impose these requirements, and as such they have intentionally created a more expansive data set.

This study is ongoing at Berkeley and is significant in that it substantiates previous studies that have determined the warming of Earth's climate. Additional information concerning the BEST study may be found at the following web site: http://berkeleyearth.org/about-data-set.

The Berkeley study was partly funded by one of the billionaire Koch brothers (who are adamantly opposed to the concept of global warming and fund multiple climate change denier efforts). The brothers own Koch Industries, a company that has chemical and fossil fuel operations. They, along with ExxonMobil, have been at the forefront of the denialist propaganda machine using the same tactics that were used by the major tobacco companies in denying that cigarette smoke causes cancer; to cast doubt on the science. If there is doubt, regulations and public support would be delayed and delay means more profits as they continue to sell their fossil fuels, become even wealthier, and pollute the planet.

Berkeley Earth became an independent non-profit 501(c) (3) in February 2013. They have been criticized for only including land station records and no ocean records and partly as a result they have a new global temperature series available. This was created by combining their land record with a Kriged ocean time series using data from HadSST3. The results fit quite well with other published series, and are closest to the new estimates by Cowtan and Way (2014). Both use HadSST3 as the ocean series and use Kriging for spatial interpolation.

### 2.7.3  The Global Historical Climatology Network

The Global Historical Climatology Network is maintained by NOAA. GHCN-Daily is an integrated database of daily climate summaries from land surface stations across the globe. Like its monthly counterpart (GHCN-Monthly), GHCN-Daily is comprised of daily climate records from numerous sources that have been integrated and subjected to a common suite of quality assurance reviews.

The number of land surface weather stations in the Global Historical Climatology Network (GHCN) has decreased in recent years. This fact is an indication of success in adding historical data. Every month data from over 1,200 stations from around the world are added to GHCN as a result of monthly reports transmitted over the Global Telecommunication System.

### 2.7.4  National Climatic Data Center

The NCDC is part of NOAA. The National Climatic Data Center (NCDC), located in Asheville, North Carolina, maintains the world's largest climate data archive and provides climatological services and data to every sector of the United States economy and to users worldwide. Records in the archive range from paleoclimatology to centuries-old journals to data less than an hour old. The Center's mission is to preserve these data and make them available to the public, business, industry, government, and researchers.

NCDC has systematically sought to increase the data holdings in the past through international projects such as the once a decade creation of World Weather Records as well as NCDC's own digitization of select Colonial Era archive data. The creation of the Global Climate Observing System (GCOS) Surface Network is one example of a specific attempt to both enhance data exchange around the world and to identify and select the "best" stations for long-term climate change purposes. The weighting scheme used to rate stations for the initial selection in the GCOS Surface Network (GSN) clearly indicates the biases climatologists have in favor of stations that have been in operation for a long time, that are rural, are agricultural research sites, and are distributed throughout the world with increasing density the farther they are away from the tropics.

Because 71 % of the world is covered by ocean, NCDC also has a strong focus on collection of observations over the World Ocean. The global ocean temperature analysis is primarily based on buoy and ship observations from the International Comprehensive Ocean Atmosphere Dataset (ICOADS), while monthly data updates come from the Global Telecommunications System (GTS). NCDC is active in a continuing multi-decadal effort to digitize historical ocean observations that contribute to ICOADS. The number of sea surface temperature observations in ICOADS has increased due to recent digitization by NCDC. Other parts of NOAA are involved in ocean buoy deployments that also contribute to ICOADS and the GTS data streams. NOAA continues to seek to increase the amount of data available for global analyses.

### 2.7.5  The Climate Research Unit at the University of East Anglia, U.K.

The Climatic Research Unit (CRU) at the University of East Anglia is widely recognized as one of the world's leading institutions concerned with the study of natural and anthropogenic climate change. The aim of the Climatic Research Unit (CRU) is to improve scientific understanding in:

- Past climate history and its impacts;
- The course and causes of climate change during the past and present centuries; and
- Prospects for future climate.

### 2.7.6  The "Hockey Stick"

Probably the most famous scientific illustration of all time in climate science was produced in a paper by Michael Mann and colleagues and reproduced by the IPCC in their 2001 assessment report. It became famous because of its shape and was immediately dubbed the "hockey stick." The original can be seen in Fig. 2.2. The

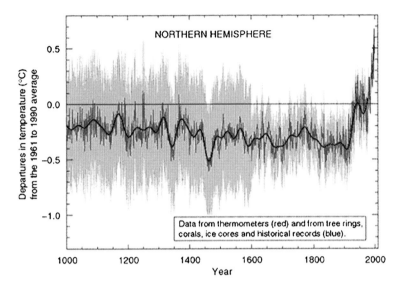

**Fig. 2.2**   The original "hockey stick" published by Mann et al. (1999)

graph shows in spectacular fashion that the warming in the late 20th century is so rapid that it may be unique, at least for the past 1,000 years.

The basic shape of the graph has been seen in several other reports, one of the most recent by Marcott et al. (2013) appearing in the prestigious journal *Science*, that shows the rate of warming over the past century is very rapid and probably unprecedented for the past 11,300 years.

## 2.8   Solar Irradiance

Solar irradiance was introduced in Sect. 1.7. Irradiance is the amount of light energy from one thing hitting a square meter of another each second. With solar irradiance, the light energy is from the Sun.

### 2.8.1   Research in Solar Irradiance

Prior to 1979 astronomers and Earth scientists did not have accurate data on the total amount of energy from the Sun that reaches the Earth's outermost atmosphere. Variable absorption of sunlight by clouds and aerosols prevented researchers from accurately measuring solar radiation before it strikes the Earth's atmosphere. The launch of the Nimbus-7 satellite in 1978 changed all that. The Earth Radiation Budget (ERB) instrument on the satellite measured levels of solar radiation just before it strikes the Earth's atmosphere.

NASA launched the Solar Radiation and Climate Experiment (SORCE) satellite on January 25, 2003. On board SORCE are four instruments that improve the accuracy of the measurements of solar energy. All instruments take readings of the Sun during each of the satellite's 15 daily orbits. The information is transmitted to ground stations at NASA's Wallops Flight Facility in Virginia and a station in Santiago, Chile. After undergoing internal review at NASA, the data are released to the public.

The Physikalisch-Meteorologisches Observatorium in Davos, Switzerland began the operational measurements of the direct solar irradiance in 1909, initiating the world's longest and still continuing time series of this kind. Research on solar irradiance is ongoing at the Observatorium at Davos.

## 2.9  Carbon Dioxide's Role in the Greenhouse Effect

Carbon dioxide's role in the greenhouse effect was introduced in Sect. 1.8 where it was seen that carbon dioxide was a very small part of the atmosphere, 0.0400 % but played a very important role in global warming. Thus it is a very important chemical compound from the standpoint of climate change science.

### 2.9.1  Research in Carbon Dioxide and Other Greenhouse Gases

Scientists know a great deal about the greenhouse gases, especially carbon dioxide. Carbon dioxide's physical properties were worked out long ago. It is a naturally occurring chemical compound composed of 2 oxygen atoms each covalently double bonded to a single carbon atom. It is a gas at standard temperature and pressure and exists in Earth's atmosphere in this state. As part of the carbon cycle, plants, algae, and cyanobacteria use light energy to photosynthesize carbohydrate from carbon dioxide and water, with oxygen produced as a waste product. Carbon dioxide is the most important of the greenhouse gases but is not the most effective.

It has been obvious since the late 18th century that the Earth's atmosphere served as a sort of blanket for keeping the planet from becoming too cold. As is often said, this is the "Goldilocks planet," not too hot or too cold, but just right for life to exist and thrive, and this is because of greenhouse gases.

The oceans play a central role in cycling carbon dioxide into and out of the atmosphere, and thus an essential role in regulating climate. There are numerous researchers who have worked on carbon dioxide and the carbon cycle. Taro Takahashi, a geochemist at Lamont-Doherty Earth Observatory, has spent the last five decades measuring the processes of the carbon cycle and deserves credit for recognizing various sinks and sources for carbon dioxide.

The U.S. National Oceanic and Atmospheric Administration (NOAA) Earth System Research Laboratory (ESRL), monitors trends in carbon dioxide. ESRL's

Mauna Loa observatory, began as part of Scripps Oceanographic Institution by Charles David Keeling in 1958, is maintained by NOAA and Keeling's son, Dr. R.F. Keeling. As of February 2014, the continuing of this facility's funding by NOAA was in jeopardy due to budget cuts.

The NOAA Atmospheric Baseline Observatories conduct long-term measurements of atmospheric gases, aerosol particles and solar radiation to provide the basis for assessing the prospects of change in the global climate and health of the atmosphere. The Arctic Atmospheric Observatory program is establishing long-term intensive measurements of clouds, radiation, aerosols, surface energy fluxes and chemistry in three different Arctic climate regimes. These measurements will be used to determine the mechanisms that drive climate change through a combination of process studies, satellite validation and modeling studies.

The main greenhouse gases are water vapor ($H_2O$), carbon dioxide ($CO_2$), methane ($CH_4$), nitrous oxide ($N_2O$), ozone ($O_3$) and the chlorofluorocarbons (CFCs). Each of these compounds, plus a few others, traps heat leaving Earth's surface (they trap heat by vibrating in the same wavelength) and re-radiate it back in the direction of the surface (they actually radiate the energy in all directions, just some of it strikes the Earth's surface directly).

Many greenhouse gases occur naturally in the atmosphere, such as carbon dioxide, methane, water vapor, ozone, and nitrous oxide, while others are synthetic. Those that are man-made include the chlorofluorocarbons (CFCs), hydrofluorocarbons (HFCs) and perfluorocarbons (PFCs), as well as sulfur hexafluoride ($SF_6$). Gases such as carbon dioxide and water vapor occur naturally in the atmosphere, and through humankind's interference with the carbon cycle (through burning forests, or mining, making cement, and burning coal, oil, and natural gas), we artificially move carbon from solid storage on or beneath the Earth (where it has been for millions of years) to its gaseous state, thereby increasing atmospheric concentrations of the gas.

Atmospheric concentrations of the greenhouse gases carbon dioxide ($CO_2$), methane ($CH_4$), and nitrous oxide ($N_2O$) in 2013 exceed the range of concentrations recorded in ice cores during the past 800,000 years.

On 10 May 2013, the concentration of climate-warming carbon dioxide in the atmosphere passed the milestone level of 400 ppm for the first time in human history. The last time so much greenhouse gas was in the air was several million years ago, when the Arctic was ice-free, savannah spread across the Sahara desert and sea level was up to 40 m higher than today.

According to the IPCC AR5 2013, with "very high confidence," the current rates of $CO_2$, $CH_4$ and $N_2O$ rise in atmospheric concentrations and the associated increases in radiative forcing are unprecedented with respect to the "highest resolution" ice core records of the last 22,000 years. There is "medium confidence" that the rate of change of the observed greenhouse gas rise is also unprecedented compared with the lower resolution records of the past 800,000 years.

The mid-Pliocene $CO_2$ concentration of between 350 and 450 ppm best compares with what we have today (400 ppm) so that the more we can learn about the conditions during that time will give us insight into what we are facing today and possibly in the near future.

The results of modeling for the IPCC AR5 WG1 report, *The Physical science Basis* (2013) indicate that on a global scale, the largest contributor to mid-Pliocene warmth is elevated $CO_2$. Global sea level was 25 m higher and the northern hemisphere ice sheet was small and short-lived before the onset of extensive glaciation over Greenland that occurred in the late Pliocene around 3 Ma (million years ago). The formation of an Arctic ice cap is signaled by an abrupt shift in oxygen isotope ratios and ice-rafted cobbles (Heinrich events) in the North Atlantic and North Pacific. Mid-latitude glaciation was probably underway before the end of the Pliocene.

The concentrations of greenhouse gases in Earth's atmosphere are of much interest to climate scientists as they are the major cause of the current warming of the planet. The Integrated Carbon Observation System (ICOS) is a new strategic research infrastructure for the European Union (EU) to quantify the greenhouse gas balance in Europe and adjacent regions. It consists of an integrated network of ecosystem long-term observation sites, a network of atmospheric greenhouse gas concentration sites and a network of ocean observations.

The ICOS project will build an infrastructure for coordinated, integrated, long-term high-quality observational data of the greenhouse balance of Europe and of adjacent key regions of Siberia and Africa. Consisting of a center for co-ordination, calibration and data in conjunction with networks of atmospheric and ecosystem observations, ICOS is designed to create the scientific backbone for a better understanding and quantification of greenhouse gas sources and sinks and their feedback with climate change.

The U.S. Environmental Protection Agency (EPA) is reducing greenhouse gas emissions by promulgating regulations for those who emit these gases. The EPA has recently (January 2014) issued proposed regulations for any new coal-fired power plants that may be built in the future. Unfortunately, they have not yet issued similar regulations for existing power plants.

- The EPA's "Inventory of U.S. Greenhouse Gas Emissions and Sinks," provides the United States' official estimate of total national-level greenhouse gas emissions. This report tracks annual U.S. greenhouse gas emissions since 1990.
- The EPA "Greenhouse Gas Reporting Program" collects and publishes emissions data from individual facilities in the United States that emit greenhouse gases in large quantities.

The main U.S. monitoring station for carbon dioxide, and for maintaining the Keeling curve, is the NOAA observatory at Mauna Loa, Hawaii founded by Charles David Keeling.

The U.K.'s Met Office, Hadley Centre on November 13, 2013 released their report on the global carbon budget.

The land and ocean typically absorb approximately half of anthropogenic emissions. Any change in the efficiency of these sinks will have significant consequences for accumulation of anthropogenic $CO_2$ in the atmosphere.

## 2.10  Carbon Dioxide and Carbon

The difference between carbon dioxide and carbon is explained in Sect. 1.9. Basically, one unit of carbon, when oxidized, produces 3.67 units of carbon dioxide.

### 2.10.1  The Carbon Cycle

The carbon cycle is the biogeochemical cycle of carbon, in its various forms, in nature. By far the most carbon within the cycle (37,000 GtC—gigatons of carbon, 1 gigaton = 1 billion tons = 1 petagram) is stored in the deep ocean waters. The main reservoirs of carbon are the deep ocean, the ocean surface, rocks of the Earth's crust, plant biomass, reactive sediments, the atmosphere, and soils. As carbon is transferred from one reservoir to another, carbon moves, or is cycled, throughout the system.

One of the world's leading scholars of the carbon cycle is Dr. David Archer of the University of Chicago. Dr. Archer is the author of several books on climate change/global warming and teaches a popular introductory course in climate science at the university and a cost-free Massive Open Online Course (MOOC). NOAA's ESRL Monitoring Division has the Global Greenhouse Gas Reference Network that monitors the movement of carbon through the cycle.

The time it takes carbon to move through the fast carbon cycle is measured in a lifespan. The fast carbon cycle is largely the movement of carbon through life forms on Earth, or the biosphere. Between 1,000 million metric tons to 100,000 million metric tons of carbon move through the fast carbon cycle every year.

Through a series of chemical reactions and tectonic (mountain-building and plate-moving) activity, carbon takes between 100 and 200 million years to move between rocks, soil, ocean, and atmosphere in the slow carbon cycle. On average, 10–100 million metric tons of carbon move through the slow carbon cycle. In comparison, human emissions of carbon dioxide to the atmosphere are on the order of 35–38 gigatons per year.

The movement of carbon from the atmosphere to the lithosphere (rocks) begins with rain. Atmospheric carbon dioxide combines with water to form a weak acid, carbonic acid that falls to the land surface every time it rains. The acid dissolves rocks (a process called chemical weathering) and releases calcium, magnesium, potassium, or sodium ions. Rivers carry the ions along with carbonic acid to the ocean as part of their dissolved load.

### 2.10.2  Current and Recent Research on Carbon

Most research on aspects of the carbon cycle is conducted by federal agencies in the U.S. and elsewhere. These agencies monitor concentrations of greenhouse gases, their emissions, sinks, and their contamination pathways. The U.S. agencies that monitor the carbon cycle are NASA, U.S. EPA, and NOAA, discussed below.

NASA research on the carbon cycle has involved, as of early 2011, two types of satellite instruments collecting information relevant to the carbon cycle. The Moderate Resolution Imaging Spectroradiometer (MODIS) instruments, flying on NASA's Terra and Aqua satellites, measure the amount of carbon in plants and phytoplankton that turn into plant matter as they grow, a measurement called "net primary productivity." The MODIS sensors also measure how many fires occur and where they burn, especially important as wildfires become larger and burn with more intensity as the ground loses moisture to the atmosphere due to global warming. Wildfires release carbon dioxide to the atmosphere.

Two Landsat satellites provide a detailed view of ocean reefs, what is growing on land, and how land cover is changing. It is possible to see the growth of a city or a transformation from forest to farm. This information is crucial because land use accounts for one-third of all human carbon dioxide emissions.

Future NASA satellites will continue these observations, and also measure carbon dioxide and methane in the atmosphere and vegetation height and structure. All of these measurements will show how the global carbon cycle is changing through time. They will help gauge the impact humans are having on the carbon cycle by releasing carbon dioxide into the atmosphere or finding ways to store it elsewhere. They will also show how the changing climate is altering the carbon cycle, and how the changing carbon cycle is altering our climate.

Most humans, however, will observe changes in the carbon cycle in a more personal way. The carbon cycle is the food we eat, the electricity in our homes, the gas in our cars, and the weather over our heads. We are a part of the carbon cycle, and so our decisions about how we live greatly affect the cycle. Likewise, changes in the carbon cycle will impact the way we live. As each of us better understands our role in the carbon cycle, we can adjust our "carbon footprint."

NOAA is responsible for acquiring and maintaining the global, regional, and local record of $CO_2$ and other greenhouse gases. NOAA's network is the backbone of the global system embraced by the World Meteorological Organization (WMO) that not only provides accurate and timely information on $CO_2$ and other gases, but also fosters international cooperation and collaboration on an issue that is of global importance. Maintaining such a network requires a dedication to accuracy, precision, cooperation, and transparency. NOAA's scientists not only are involved in maintaining a large portion of the worldwide network, but also provide the calibration necessary for an integrated network and serve on several advisory groups and expert committees for assuring quality control, improving understanding, and identifying future needs.

U.S. EPA uses national energy data, data on national agricultural activities, and other national statistics to provide a comprehensive accounting of total greenhouse gas emissions for all man-made sources in the United States. EPA is one of the agencies involved in the U.S. CLIVAR program.

U.S. CLIVAR (Climate Variability and Predictability Program) has released a new Science Plan outlining its research goals and strategies for the next 15 years. U.S. CLIVAR is made up of thirteen federal agencies. U.S. CLIVAR is a national research program investigating the variability and predictability of the global climate system on time scales ranging from seasons to hundreds of years, with a particular emphasis on the role of the ocean.

### 2.10.3  The Enhanced Greenhouse Effect

The enhanced greenhouse effect was defined in Sect. 1.9.1. It is caused by greenhouse gases added to the atmosphere by humankind. The main gas added by humans is carbon dioxide and this is mainly due to the combustion of fossil fuels. Humans have been taking materials from the Earth that were formed millions of years ago, stored as materials within the Earth, and adding it to the fast carbon cycle on perhaps an unprecedented scale in the history of the planet.

Humans had no idea that the concentration of a greenhouse gas was dramatically increasing in the atmosphere until the Keeling Curve was produced that showed carbon dioxide concentrations continually increasing since 1958. This was a startling discovery!

### 2.10.4  Climate Sensitivity Research

Climate sensitivity was discussed in Sect. 1.9.4 of this text and research is emphasized in this section. Climate sensitivity is a metric used to characterize the response of the global climate system to a given forcing. (An analogous way to think of climate sensitivity is to consider poking a sleeping bear with a sharp stick and seeing what his reaction will be. The bear's reaction (sensitivity) is a great deal more predictable than is climate sensitivity.)

Climate sensitivity specifically due to $CO_2$ is often expressed as the temperature change in degrees Celsius associated with a doubling of the concentration of carbon dioxide in Earth's atmosphere, but climate is sensitive to much more complex forcings and feedbacks than just $CO_2$. There is a great deal of uncertainty associated with climate sensitivity and with both feedbacks and forcings.

Climate scientists often mean different things when they talk about climate sensitivity, so terms need to be well-defined. The equilibrium climate sensitivity (ECS) refers to the equilibrium change in global mean near-surface air temperature that would result from a sustained doubling of the atmospheric equivalent of carbon dioxide concentration, commonly expressed as $CO_{2\text{-eq}}$.

The IPCC Fourth Assessment Report (AR4) stated that climate sensitivity to a doubling of $CO_2$ was likely to be in the range of 2–4.5 °C with a best estimate of about 3 °C, and is very unlikely to be less than 1.5 °C. Values substantially higher than 4.5 °C cannot be excluded, but agreement of models with observations is not as good for those values. This is a change from the IPCC Third Assessment Report (TAR), which said it was "likely to be in the range of 1.5–4.5 °C." In the IPCC AR5 this was changed back to 1.5–4.5 °C. Figure 2.3 shows sensitivity results

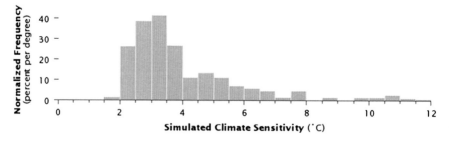

**Fig. 2.3** This image shows a frequency distribution of climate sensitivity, based on model simulations. To understand how uncertainty about the underlying physics of the climate system affects climate predictions, scientists have a common test: they have a model predict what the average surface temperature would be if atmospheric carbon dioxide concentrations were to double pre-industrial levels (NASA, public domain)

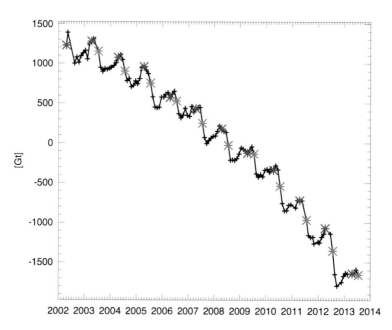

**Fig. 2.4** Monthly changes in the total mass (in gigatonnes) of the Greenland ice sheet estimated from GRACE measurements since 2002. The *blue* and *orange asterisks* denote April and July values, respectively (NOAA/NSIDC, public domain)

of modeling studies and it is apparent that they cluster around 3 °C, which is the same result as studies based on paleoclimate, climate models, and empirical data (Fig. 2.4).

In a paper published 16 September 2013 by Jim Hansen and colleagues, in the Philosophical Transactions of the Royal Society (A), entitled "Climate sensitivity, sea level and atmospheric carbon dioxide." They define climate sensitivity as follows:

"Climate sensitivity (S) is the equilibrium global surface temperature change ($\Delta$Teq) in response to a specified unit forcing (F) after the planet has come back to energy balance, the formula for climate sensitivity (below),

$$S = \Delta Teq/F$$

is the eventual (equilibrium) global temperature change per unit forcing. Climate sensitivity depends upon the condition of the climate at the start of the forcing, climate feedbacks, and the many physical processes that come into play as climate changes in response to a forcing. Positive (amplifying) feedbacks increase the climate response, whereas negative (diminishing) feedbacks reduce the response."

As we have seen, climate sensitivity partially depends on the initial climate state and should be able to be accurately inferred from precise paleoclimate data. Pleistocene (see Appendix A) climate changes from glacial-interglacial stages yield a fast-feedback climate sensitivity of $3 \pm 1$ °C for a 4 W/m$^2$ $CO_2$ forcing if Holocene warming relative to the Last Glacial Maximum (LGM) is used as calibration, but the error (uncertainty) is substantial and partly subjective because of poorly defined LGM global temperature and human influences in the Holocene (Anthropocene).

Climate sensitivity in terms of a global mean temperature response to a 4 W/m$^2$ $CO_2$ forcing is an alternative to defining sensitivity as the climate's temperature response to a doubling of $CO_2$. This introduces a standard forcing into the equation. Feedbacks to the forcing represent the main problem with determining the sensitivity; but the temperature arrived at by several methods is close to 3 °C.

Hansen et al. 2013 point out that humankind is now the dominant force driving changes in the Earth's atmospheric composition and climate. The largest climate forcing today is the human-made increase in atmospheric greenhouse gases (GHGs), especially $CO_2$ from the burning of fossil fuels. This point is beyond dispute and can now be considered a fact although some still argue the facts.

## 2.10.5  Carbon Capture and Sequestration or Carbon Capture and Storage

Carbon Capture and Sequestration or Carbon Capture and Storage (CCS) was introduced in Sect. 1.9.5. With all CCS schemes, it is paramount that $CO_2$ can never escape to the atmosphere. A $CO_2$ leak would defeat the whole purpose of the concept of CCS which is to remove carbon dioxide from the fast carbon cycle so that its concentration in the atmosphere can be reduced to 350 ppm and stay there. There are several pilot programs ongoing in countries throughout the world at present. Some of these operations are for temporary storage of $CO_2$ to be later used for enhanced oil recovery (EOR) and will not meet the specifications for sequestration.

As of September 2012, the Global CCS Institute identified 75 large-scale integrated projects in its 2012 Global Status of CCS report. As of February 2014 there are 21 large-scale projects in operation or construction—a 50 % increase since

2011. These have the capacity to store up to 40 million tonnes of $CO_2$ per annum, equivalent to 8 million cars being taken off the road.

Six projects, with a combined capture capacity of 10 million tonnes of $CO_2$ per annum, are in advanced stages of development planning and may make a final investment decision during 2014. *The Global Status of CCS February 2014* report was recently published and is available from the following website: http://www.glo balccsinstitute.com/get-involved/in-focus/2014/02/global-status-ccs-february-2014.

There are many more of these storage facilities throughout the world related to the oil and gas industry. Some may be candidates for CCS but many will not be. Having CCS facilities related to oil and gas companies is like "having the fox guard the henhouse."

### *2.10.6  Tipping Points*

A tipping point may be a temperature range; a high, average, or low temperature; a water, land, or air chemical or species. In fact, a tipping point is a point at or beyond which there is no hope for reversal. If humankind has allowed the global temperature to rise at or beyond the point of melting the Greenland ice sheet, which is becoming more apparent as the years pass and the ice cap is still melting, then we have already reached and exceeded the tipping point for the Greenland ice sheet.

There are a number of researchers studying the Greenland ice to try and determine just where it is in the melting process and just what the processes are, how they are working, and if a "tipping point" has been reached. The greatest melt season in Greenland's history occurred during 2012. The melt season of 2013 was not as great as in 2012; it was still substantial, but of course there was less ice to melt in 2013.

Two prominent researchers on the Greenland ice sheet, among many, that are very concerned about the tipping point are Richard Alley and Jason Box. Dr. Richard Alley has been at the forefront of the Greenland ice research. Dr. Alley is the Evan Pugh Professor of Geosciences at Penn State University, a prolific author, and testifier to committees of the U.S. Congress. He is the author of *Earth: The Operator's Manual*, *The Two–Mile Time Machine*, and teaches a MOOC on climate change science.

Dr. Jason Box, Geological Survey of Denmark and Greenland (GEUS), has been investigating Greenland ice sheet sensitivity to weather and climate as part of 23 expeditions to Greenland since 1994. Box has authored or co-authored over 50 peer-reviewed publications related to Greenland-cryosphere-climate interactions.

## 2.11  Climate Forcing and Climate Feedbacks

Climate forcings and feedbacks were introduced in Sect. 1.10. Climate forcing is something that causes the global climate to change. Feedbacks either enhance or retard a forcing.

## 2.11.1 Research into Climate Forcing and Climate Feedbacks

James Hansen and his colleagues at NASA/GISS have been the research leaders in both climate forcings and feedbacks for some time during the late 20th and into the 21st century. Many of their publications are given at the end of this chapter.

## 2.12 Climate Models

Climate models were introduced in Sect. 1.11. There are different types of climate models and many different climate modelers. One of the founders of modern climate modeling is Syukuro "Suki" Manabe, a Japanese-born meteorologist and climatologist who worked at NOAA's Geophysical Fluid Dynamics Laboratory (GFDL), first in Washington, DC and later in Princeton, New Jersey, who pioneered the use of computers to simulate global climate change and natural climate variations.

In 1967 Manabe and Richard Wetherald demonstrated that increasing atmospheric carbon dioxide concentrations would increase the altitude at which Earth radiated heat to space. In 1969 Manabe and Kirk Bryan published the first simulations of the climate of a planet with coupled ocean and atmosphere models, establishing the role of oceanic heat transport in determining global climate. Throughout the 1970s and 1980s Manabe's research group published seminal papers using these models to explore the sensitivity of Earth's climate to changing greenhouse gas concentrations. These papers formed a major part of the first global assessments of climate change published by the IPCC.

The more complex climate models consist of differential equations representing the processes occurring on land, in the sea, and in the atmosphere. These models simulate the intricacies of Earth's climate and allow climate scientists to conduct "what if" experiments, a rarity in the Earth sciences (we don't have another Earth with which to experiment). Figure 2.5 shows some of the input to various climate models (identified by their acronyms).

The most complex models used to simulate the climate are General Circulation Models or General Climate Models (GCMs). In using a climate model, a climate scientist first makes a "control run" to verify if known features of the climate system, such as ocean and atmosphere currents, behave the same as they do over time based on empirical evidence.

Then the climate model is run in changing conditions, simulating the last couple of centuries using the best estimates of the climate "forcings" (or drivers of climate change) at work over that time period. These forcings include solar activity, volcanic eruptions, changing greenhouse gas concentrations, and human modifications of the landscape (such as deforestation, spread of urban areas, etc.).

The results are then compared to actual observations of things like global temperatures, local temperatures, and precipitation patterns. Did the model capture the big picture and how did it handle the finer details? Which fine details did it simulate poorly and why?

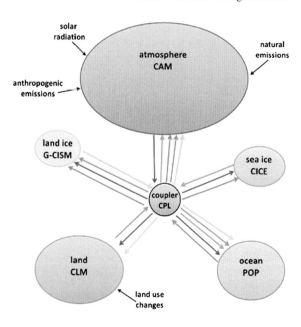

**Fig. 2.5** Diagram of software architecture for the community earth system model. Coupled models use interacting components simulating different parts of the climate system. *Bubble size* represents the number of lines of code in each component of this particular model (*Source* Kaitlin Alexander and Steve Easterbrook, from Skepticals cience.com)

Climate models are living scientific tools that are constantly evolving rather than pieces of software built to achieve a certain goal. The models are tested and streamlined to give the best results possible. Many trial runs are completed before the model is ready to be used in climate simulations.

An example of model testing is by using proxy records of climate from cores of ice or ocean sediments that are limited to providing information about the geographic area from which they were collected (see Fig. 2.6). Climate models can help fill in the rest of the global picture. A model simulation of actual events like an immense ice-dammed lake draining into the North Atlantic and disrupting ocean circulation can be compared to a network of proxy records to see if the simulated climate impact is consistent with what the proxies show. If the match is poor, then perhaps the observed change in climate was caused by something else, or perhaps the model needs further tweaking.

Models of actual events or trends are never perfect but they represent one of the best tools available to climate scientists, especially for projecting climate conditions that might be expected in the future. It is the only means by which climate scientists can experiment with the climate.

### 2.12.1 Current and Recent Research

The majority of GCM climate modeling occurs at national and international research laboratories that have or can access large computing facilities necessary to run the models. Examples include the National Center for Atmospheric Research (NCAR,

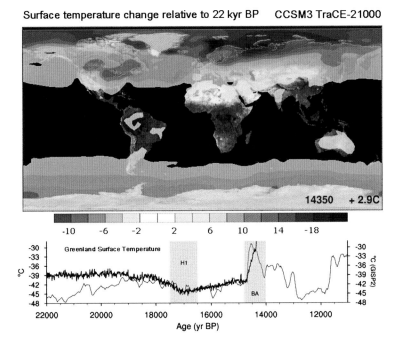

**Fig. 2.6** Snapshot from a modeling experiment simulating the last 22,000 years. In the graph at the *bottom*, the *dark line* represents simulated surface temperature over Greenland and the *lighter line* shows data from a Greenland ice core (from NCAR, National Center for Atmospheric Research/University Corporation for Atmospheric Research, public domain)

in Boulder, Colorado, USA), the Geophysical Fluid Dynamics Laboratory (GFDL, in Princeton, New Jersey, USA), NASA/GISS, NOAA, Los Alamos National Laboratory, Lawrence Livermore, Oak Ridge and other National Laboratories in the U.S., the Hadley Centre for Climate Prediction and Research (in Exeter, UK), the Max Planck Institute for Meteorology in Hamburg, Germany, and the Laboratoire des Sciences du Climat et de l'Environnement (LSCE), France, to name but a few.

Working with large entities such as the U.S. government or the international modeling community, one has to become familiar with the "alphabet soup" of multiple acronyms such as those of the following groups: Under the World Climate Research Programme (UWCRP) the Working Group on Coupled Modeling (WGCM) established the Coupled Model Intercomparison Project (CMIP) as a standard experimental protocol for studying the output of coupled atmosphere-ocean general circulation models (AOGCMs).

CMIP provides a community-based infrastructure in support of climate model diagnosis, validation, intercomparison, documentation and data access. Virtually the entire international climate modeling community has participated in this project since its inception in 1995. The Program for Climate Model Diagnosis and Intercomparison (PCMDI) archives much of the CMIP data and provides other support for CMIP. PCMDI's CMIP effort is funded by the Regional and Global

Climate Modeling (RGCM) Program of the Climate and Environmental Sciences Division of the U.S. Department of Energy's Office of Science, Biological and Environmental Research (BER) program.

Coupled atmosphere-ocean general circulation models allow the simulated climate to adjust to changes in climate forcing, such as increasing atmospheric carbon dioxide. CMIP began by collecting output from model "control runs" in which climate forcing is held constant. Later versions of CMIP have collected output from an idealized scenario of global warming, with atmospheric $CO_2$ increasing at the rate of 1 % per year until it doubles at about year 70. CMIP output is available for study by approved diagnostic sub-projects.

Climate modeling activity is extensive both in the United States and internationally. Climate models have advanced over the decades to become capable of providing much useful information that can be used for decision making. Critics of climate modeling do not understand the concept.

Phase three of CMIP (CMIP3) included "realistic" scenarios for both past and present climate forcing. The research based on this dataset provided much of the new material underlying the IPCC Fourth Assessment Report (AR4). CMIP5 is the most recent version and is being used for IPCC AR5.

The DOE-funded Program for Climate Model Diagnosis and Intercomparison (PCMDI) at Lawrence Livermore National Laboratory in California has been instrumental in developing CMIP, including archiving, analysis, and quality control of model output, although CMIP now has broad international institutional support. CMIP has developed into a vital community-based infrastructure in support of climate model diagnosis, validation, intercomparison, documentation, and data access.

Dr. Ben Santer, a Research Scientist for the Program for Climate Model Diagnosis and Intercomparison at Lawrence Livermore National Laboratory in California is one of the world's leading climate researchers and has been instrumental in the development and use of the PCMDI for the modeling community. Unfortunately, Dr. Santer has also been brutally targeted by some of the more aggressive climate change deniers.

## 2.13  Earth's Atmosphere

Earth's atmosphere was introduced in Sect. 1.12. Research in atmospheric processes related to climate change is being done by climate scientists (climatologists and others), meteorologists, physicists, geochemists, and atmospheric scientists of all kinds. The parts and processes of the atmosphere are fairly well known but mysteries still abound.

A few of the outstanding researchers in atmospheric sciences were recently recognized by the American Geophysical Union (AGU) and are listed below:

- Cecilia Bitz (University of Washington): For advancing our ability to model climate in numerous ways, especially relating to sea ice;
- Paul Ginoux (NOAA GFDL): For sustained pioneering work on aerosols;

- Mark Jacobson (Stanford University): For his dominating role in the development of models to identify the role of black carbon in climate change;
- Sergey Nizkorodov (University of California, Irvine): For elucidating at the molecular level the formation, growth and reactions of organic molecules in the atmosphere;
- Ping Yang (Texas A&M University): For fundamental research in radiative transfer and remote sensing.

Dr. Richard C.J. Somerville is a climate scientist who is a Distinguished Professor Emeritus at Scripps Institution of Oceanography, University of California, San Diego, U.S. His research is focused on critical physical processes in the climate system, especially the role of clouds and the important feedbacks that can occur as clouds change with a changing climate. The influence of clouds on global warming is one of the difficulties with climate models. He has authored over 200 peer-reviewed publications in meteorology and climatology. He is part of a team from U.C. San Diego teaching a MOOC on climate change. There are of course many other outstanding researchers involved with atmospheric processes that are too numerous to mention.

## 2.14 Earth's Land Surface

The land surface and climate are related mainly through the topography of the land and the latitude at which the land is located. Climate change is more than just warming or cooling of Earth but involves wind, rainfall, snowfall, ice storms, and all other things that we call weather. Weather is short term, climate is long term.

Most carbon in the terrestrial biosphere is stored in forests: they hold 86 % of the planet's terrestrial above-ground carbon and forest soils also hold 73 % of the planet's soil carbon.

When Earth is in the throes of climate change, the weather is also affected. For example, the warming Earth is rapidly removing sea ice from the Arctic Ocean. As Arctic sea ice becomes less, darker ocean water is exposed to Sunlight and it absorbs more of the Sun's energy that was previously reflected by the sea ice. This begins a feedback loop. Less ice causes more warming which causes more ice to melt and so on. The same thing happens to glaciers on land; a feedback loop develops.

Currently, people in the Northern Hemisphere are experiencing what most would deem strange weather; extreme drought in the southwestern U.S., extreme cold and snowfall in the mid-western and northeastern U.S., extreme rainfall and flooding in the British Isles, etc.; and this is all because of climate change/global warming.

Areas with topography consisting of high mountain ranges affect climates in the Rockies, Alps, Andes, and Himalayas. Climates in these areas tend to be colder than in surrounding lowlands. Low-lying areas in mid-latitudes tend to have mild but often changing climate. In some mountainous areas, rain shadows develop on the leeward side of mountains causing a dry climate on the leeward side.

Research in the topics of land, landscapes, and land climates are mainly in the domains of geologists, geographers, and ecologists. The movement of $CO_2$ from the land biosphere to the atmosphere is an active area of research. On November

21 2013 the U.S. Global Carbon Project released the *Global Carbon Budget 2013*
and the new *Global Carbon Atlas*, an online interactive platform with distinctive
tools to deliver carbon information to educators and the general public at the following website: http://www.globalcarbonatlas.org.

## 2.15  The World Ocean

The World Ocean was introduced in Sect. 1.14. Warming has been detected to a
depth of 2,000 m with the majority of warming from the ocean "skin" at the surface to a depth of 700 m with more recent warming at greater depths. Warming
of the ocean will continue as long as Earth has an energy imbalance and for
hundreds of years after equilibrium is attained. Today, ocean temperature data
are obtained mainly by over 3,600 Argo floats scattered around the world. Argo
float data are gathered and studied by scientists at NOAA. Argo is a major contributor to the World Climate Research Program's (WCRP's) Climate Variability
and Predictability Experiment (CLIVAR) project and to the Global Ocean Data
Assimilation Experiment (GODAE). The Argo array is part of the Global Climate
Observing System/Global Ocean Observing System (GCOS/GOOS).

## 2.16  Atmosphere-Land Surface-Ocean Interfaces

The World Ocean plays an important role in regulating the amount of $CO_2$ in the
atmosphere because $CO_2$ can move quickly into and out of the ocean. Once in the
ocean, the $CO_2$ no longer traps heat. $CO_2$ also moves quickly between the atmosphere and the land biosphere (plants and animals that are living on land).

Of the three places where carbon is stored (atmosphere, ocean, and land biosphere) approximately 93 % of the $CO_2$ is found in the ocean. Approximately
90–100 Pg of carbon moves back and forth between the atmosphere and the ocean,
and between the atmosphere and the land biosphere.

Sea water and $CO_2$ mix much more slowly than the atmosphere and $CO_2$, so
there are large horizontal and vertical changes in $CO_2$ concentration in the ocean.
In general, tropical waters release $CO_2$ to the atmosphere, whereas high-latitude
waters take up $CO_2$ from the atmosphere. $CO_2$ is also about 10 % higher in the
deep ocean than at the surface.

### 2.16.1  Current and Recent Research

The World Ocean serves humankind as a source of food, to moderate climate,
to store minerals, to provide oil and gas from rocks under the sea, to give rise to
clouds, to circulate ocean waters, and to provide moisture for rainfall. There is
ongoing research in all aspects of oceanography.

There are several oceanographic institutions that study and actively do research in ocean waters of the world. In the U.S., the National Oceanographic and Atmospheric Administration (NOAA) is the lead federal agency conducting oceanographic research. NASA also contributes to oceanic research with instrumentation on numerous satellites.

Each coastal state has one or more oceanographic institutions. There are also universities and colleges with departments of, or courses in, oceanography that do oceanographic research. Woods Hole Oceanographic Institution on the U.S. east coast and Scripps Institution of Oceanography on the west coast are two of the better known research centers. Japan has the Japan Agency for Marine-Earth Science and Technology and the Research Institute for Global Change (RIGC) that has a large contingent of science researchers in global climate science. Australia has the Commonwealth Scientific and Industrial Research Organisation (CSIRO). Canada has the Institute of Ocean Sciences and the Oceanography Department, Dalhousie University, Halifax, Nova Scotia. China has the Institute of Ocean Sciences, Swire Institute of Marine Science, The University of Hong Kong, and the Institute of Oceanology, Chinese Academy of Sciences among others.

In Britain, there is a major research institution: the National Oceanography Centre, Southampton which is the successor to the Institute of Oceanography. University of Exeter in the U.K. does research on climate change and ocean acidification. In Australia, CSIRO Marine and Atmospheric Research, known as CMAR, is a leading center. In 1921 the International Hydrographic Bureau (IHB) was formed in Monaco. Japan also has numerous research centers for oceanography; The Japanese National Oceanographic Data Center archives and disseminates marine data obtained from various Japanese marine research institutes.

Obtaining temperature information from ocean waters has been greatly increased by the Argo Program. Prior to implementing the program, data were obtained by a variety of methods with varying degrees of reliability. The Argo program is uniformly collecting temperature and salinity data from the World Ocean to depths of 2,000 m.

The Argo Program was conceived and designed by a group of U.S. academic and NOAA scientists to take data from the altimetry instruments aboard NASA's Jason and TOPEX/Poseidon satellites that measure differences in ocean surface elevation and combine them with ocean temperature, salinity, and current profiles gathered by Argo floats. The program builds on the ocean sampling method developed during the World Ocean Circulation Experiment (WOCE) of the 1990s by creating a global array of over 3,600 profiling floats spaced roughly in three-degree grids throughout ice-free ocean areas.

WOCE employed radar satellites that continuously monitored the shape of the ocean surface to improve understanding of ocean features and as validation for ocean models. The data collected during WOCE were also used to improve existing ocean models used for global climate prediction. Operational uses included assimilating temperature and salinity data into global ocean models. Battery-operated, neutrally buoyant floats that were the forerunners of today's Argo float collected these data. These early instruments stayed at a specified depth and surfaced every few weeks so that satellites could monitor their position and determine subsurface currents. These early floats were then modified to take temperature and salinity profiles (NOAA 2013).

Argo floats are deployed at the sea surface. The floats sink to a prescribed depth (usually 1,000–2,000 m, sometimes deeper), collecting data on the descent. They rest at depth for a prescribed period and rise back to the surface, collecting data as they rise. Some floats are designed to sink to 2,000 m before ascending to the surface. At the surface, the floats transmit to satellites the positioning and oceanographic data collected since their last transmission. The floats then sink again to repeat the cycle that typically lasts 10 days. Over the course of an average lifetime, an Argo float completes about 150 cycles.

Currently, more than 60 Argo floats are measuring dissolved oxygen using two types of sensors that have returned stable results for over a year. These data may allow scientists to estimate ocean uptake of carbon dioxide, an important piece of the puzzle in understanding the fate of one of the main greenhouse gases ($CO_2$) that we've seen as a forcing agent in climate change.

Decreasing oxygen levels in the ocean caused by climate change and agricultural (mainly nitrogen) runoff, combined with other chemical pollution and over-fishing are undermining the ability of the ocean to withstand so-called "carbon perturbations," meaning its role as Earth's "buffer" is seriously compromised. As the ocean goes, so goes all life on Earth.

Springer Publishers (www.springer.com) publishes the Journal of Oceanography. There are several other journals that publish original research on oceanography and ocean contributions to climate change, namely Geophysical Research Letters, Science, and Nature.

Shakhova et al. (2013) point out that ocean bottom water temperatures are increasing more than had been recognized, in particular in near-coastal shallow waters. Stefan Rahmstorf, mentioned previously, is currently one of the leading students of the World Ocean and climate science in general. He is a professor at the Potsdam Institute for Climate Impact Research in Germany. His work focuses on the role of ocean currents in climate change, past and present and he publishes papers on rising sea level. He is a frequent contributor to the website www.realclimate.org, which is a highly recommended site for accurate climate change information.

### 2.16.2  Ocean Acidification

Ocean acidification was introduced in Sect. 1.16 where it was defined as a reduction in pH. It doesn't mean that the ocean is acidic. Actually, most ocean water has a pH of 8.1, well on the basic side of neutral which is 7.

### 2.16.3  Ocean Acidification Research

There is little doubt that ocean acidification is caused by the rising concentration of $CO_2$ in the atmosphere. Ocean waters don't mix as rapidly as does the atmosphere so that $CO_2$ is not dispersed evenly throughout the World Ocean. There is currently a great deal of research being conducted on the behavior of $CO_2$ in marine waters, mainly near the ocean surface in studying the ocean-atmosphere interface gas exchanges. There is also abundant research on the reduction in pH and its effects on marine life.

Trends toward lower pH can now be clearly seen in sustained observations recorded at ocean time series stations and during repeated geochemical surveys. In coming decades, ocean acidification could affect some of the most fundamental biological and geochemical processes of the sea.

Ocean Carbon and Biogeochemistry (OCB) was established in 2006 as one of the major activities of the U.S. Carbon Cycle Science Program, an interagency body that coordinates and facilitates activities relevant to carbon cycle science, climate, and global change issues. The scientific mission of OCB is to study the evolving role of the ocean in the global carbon cycle, in the face of environmental variability and change through studies of marine biogeochemical cycles and associated ecosystems.

Rutgers University, the State University of New Jersey, has an active research program in ocean acidification as do Woods Hole and Scripps oceanographic institutions. NOAA is the lead U.S. government agency for studies involving the ocean.

### 2.16.4 The "Pause" in Surface Temperature

Since 1998 there has been a slowdown of the temperature rise in the lower atmosphere at Earth's surface. Climate change deniers have stated that this shows no warming since 1998. There has been plenty of warming since 1998 as discussed below. The rate of warming has actually increased since 1998. We know that the Earth is receiving more energy than it is losing because of the energy imbalance, so the heat has to be going somewhere.

A study by Cowtan and Way (2014), British (Cowtan) and Canadian (Way) researchers, shows that the global temperature rise since 1997–1998 has been greatly underestimated. The reason is because of the data gaps in the weather station network, especially in the Arctic but also in Antarctica and Africa. If these data gaps are filled using satellite measurements and Kriging (a statistical method), the warming trend is more than doubled in the widely used HadCRUT4 data, and the much-discussed "warming pause" has disappeared.

The warmest-year rankings is also evidence that there has been no pause. Of the 11 warmest years on record, 10 are since 1998. It is obvious that there has been no pause (Table 2.1).

Several studies have shown that much of the excess heating of the planet due to the radiative imbalance from ever-increasing greenhouses gases has gone into the ocean, rather than the atmosphere. In a new paper, England et al. (2014) show that this increased ocean heat uptake, which has occurred mostly in the tropical Pacific, is associated with an anomalous strengthening of the trade winds. Stronger trade winds push warm surface water toward the west, and bring cold deeper waters

**Table 2.1** The 11 warmest years on record

| Rank − 1 = Warmest | Years | Anomaly (°C) | Anomaly (°F) |
|---|---|---|---|
| 1 | 2010 | 0.66 | 1.19 |
| 2 | 2005 | 0.65 | 1.17 |
| 3 | 1998 | 0.63 | 1.13 |
| 4 (tie)[a] | 2013 | 0.62 | 1.12 |
| 4 (tie)[a] | 2003 | 0.62 | 1.12 |
| 6 | 2002 | 0.61 | 1.10 |
| 7 | 2006 | 0.60 | 1.08 |
| 8 (tie)[a] | 2009 | 0.59 | 1.07 |
| 8 (tie)[a] | 2007 | 0.59 | 1.06 |
| 10 (tie)[a] | 2004 | 0.57 | 1.04 |
| 10 (tie)[a] | 2012 | 0.57 | 1.03 |

Period of record: 1880–2013
[a]*Note* Tie is based on temperature anomaly in °C
The table is drawn from NOAA National Climatic Data Center, State of the Climate: Global Analysis for Annual 2013, published online December 2013, retrieved on February 19, 2014 from http://www.ncdc.noaa.gov/sotc/global/2013/13

to the surface to replace them. This happens every time there is a major La Niña event, which is why it is globally cooler during La Niña years.

Santer et al. (2014) in *Nature Geoscience* suggest that 17 small volcanic eruptions since 1999 have had a modest cooling effect, and that incorporating them into models helps explain divergences in temperatures of the lower part of the atmosphere. Both papers England et al. (2014) and Santer et al. (2014) suggest that the factors behind the hiatus are temporary, and do not undermine longer-term projected warming.

## 2.17  The Cryosphere—Current and Recent Research

The cryosphere was introduced in Sect. 1.19 and everything of which it was comprised was defined. This section will concentrate on research, both recent and current concerning the cryosphere.

Glacial research is conducted by federal governments worldwide, but especially in areas where active glaciers are still present, and also by public and private institutions. One of the foremost glacial programs in the U.S. is at the Ohio State University (OSU) in Columbus, Ohio. OSU has the Byrd Polar Research Center (BPRC). Research at the Center focuses on the role of cold regions in the Earth's overall climate system, and encompasses geological sciences, geochemistry, glaciology, paleoclimatology, meteorology, remote sensing, ocean dynamics, and the history of polar exploration. Much of their research is supported by the U.S. National Science Foundation (NSF). The NSF supports a great deal of cryosphere research especially in the Arctic and Antarctic.

After the Arctic and the Antarctic, the Tibetan Plateau and the Himalayas have Earth's largest store of ice, in more than 46,000 glaciers (many of which are disappearing) and vast expanses of permafrost. It is much less studied than its high-latitude counterparts, even though many more lives depend on it. It has been referred to as Earth's "Third Pole."

The Tibetan Plateau, the "Third Pole" is also known as "Asia's water tower" because its glaciers feed the continent's largest rivers, which sustain 1.5–2 billion people across ten countries. These glaciers are melting fast, filling lakes that can overflow and flood valleys. Yet little is known about how climate change is unfolding there. To attempt to rectify this, Third Pole Environment (TPE), an international program led by the Chinese Academy of Sciences' Institute of Tibetan Plateau Research (ITP) in Beijing, has held workshops for researchers in the region to lay plans to fill the knowledge gap, and discuss findings that add to the urgency. Other countries that have regions affected by the melting ice are also beginning to start research projects at the "third pole."

The U.S. Geological Survey (USGS) studies and monitors glaciers throughout the world. They have monitoring stations at the U.S. glacial sites (Glacier National Park, the Cascade Mountains, and Rocky Mountain National Park).

Glaciers in the Canadian Rockies are studied by the Geological Survey of Canada and by Canadian Province geological surveys. Natural Resources Canada is Canada's foremost geomatics (geospatial technologies) and geoscience organization. Additional information on its mission and functions can be gathered from the following website: http://www.nrcan.gc.ca/earth-sciences.

The British Geological Survey (BGS) operates an observatory site at Virkisjökull Glacier in south-east Iceland, studying the evolution of the glacier and the surrounding landscape and their responses to regional climate. Sensors at the site are constantly collecting climate and seismic data. Repeated high resolution surveys study how both the glacier and land surface, and the deposits beneath, change over time.

Virkisjökull is retreating rapidly, like most glaciers in Iceland and around the world. Since 1996, the glacier margin has retreated nearly 500 m.

The BGS interpret the combined results of constant monitoring, using automatic systems, plus field surveying of land-surface change, glacier hydrology, and evolution of the shallow subsurface at the Virkisjökull observatory.

## 2.17.1  Sea Ice

Sea ice develops when ocean waters reach their freezing point. It doesn't change sea level when it forms or melts because it is formed from sea water. Sea ice forms in polar regions, both the North and South Poles and it affects Earth's albedo.

Sea ice in the Northern Hemisphere has been getting thinner and there is less and less of it since satellite records were first started in 1979. It is a critical component of our planet because it influences climate, wildlife, and people who live in the Arctic. It forms during the winter and melts during the summer. Water below sea ice has a higher concentration of salt and is denser than surrounding ocean water, so it sinks. In this way, sea ice contributes to the ocean's global "conveyor-belt" circulation. Cold, dense, polar water sinks and moves along the ocean bottom toward the equator, while warm water from mid-depth to the surface travels from the equator toward the poles. Changes in the amount of sea ice can disrupt normal ocean circulation leading to changes in global climate.

When sea ice melts, less Sunlight is reflected and the darker ocean waters absorb more heat. This is the beginning of a feedback loop, as more ocean is heated it causes more sea ice to melt which causes more ocean heating and so on.

The U.S. National Snow and Ice Data Center's (NSIDC) research and scientific data management activities are supported by NASA, the National Science Foundation (NSF), the National Oceanic and Atmospheric Administration (NOAA), and other federal agencies, through competitive grants and contracts. More information about NSIDC can be found at the following website: http://nsidc.org/cryosphere/.

## 2.17.2  Current and Recent Research

The Arctic is still warming at a faster rate than the rest of the planet. The Arctic continued to shift to a warmer, greener state in 2013. That was the headline from the latest "Arctic Report Card," an annual update prepared by scientists from the National Oceanic and Atmospheric Administration (NOAA) and partner organizations such as NASA. Scientists are tracking a variety of environmental indicators, including air temperature, snow cover, sea ice extent, ocean temperature, vegetation growth, and wildlife behavior. In comparison to 2012, most indicators in 2013 were closer to their long-term averages, but signs of change (fueled by long-term warming) were still present.

Most surface waters within the Arctic Circle were warmer than average in summer 2013. Figure 2.7 shows sea ice extent on September 13, 2013. Sea ice extent is solid white. Although some areas experienced unusually cool sea surface temperatures (SSTs) in August 2013, especially in the Chukchi and East Siberian Seas, unusually high temperatures dominated most of the Arctic Ocean and surrounding straits and seas.

Warm waters in the eastern Arctic were probably related to an earlier-than-normal retreat of sea ice from the area and possibly an inflow of warmer water from the North Atlantic. Retreating sea ice would have left the Kara and Barents Seas exposed to warm summer Sunlight. Meanwhile, on the western side of the Arctic, sea ice retreat was later and less extensive than normal, contributing to cooler-than-average surface temperatures in the Chukchi and East Siberian Seas.

By September, surface waters around the Barents Sea Opening (between Svalbard and Scandinavia) were about 5 °F (3 °C) warmer than they were in 2012. Southern Barents Sea temperatures reached 52 °F (11 °C), which is 9 °F (5 °C) warmer than the 1977–2006 average.

Sea ice extent is not the same as the amount of sea ice present. The extent of sea ice is the area covered by sea ice. Climate scientists know that Arctic sea ice is losing volume and this continues year after year. The thinner sheets of sea ice

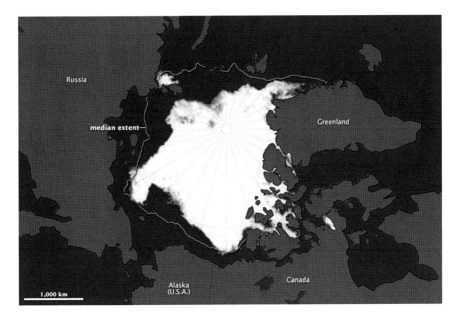

**Fig. 2.7**   After an unusually cool summer in the northernmost latitudes, Arctic sea ice appears to have reached its annual minimum extent on September 13, 2013. Analysis of satellite data by the National Snow and Ice Data Center (*NSIDC*) showed that sea ice extent shrunk to 5.10 million square kilometers (1.97 million square miles). The extent of sea ice this September is substantially greater than last year's record low. On September 16, 2012, Arctic sea ice spread across just 3.41 million square kilometers (1.32 million square miles)—the smallest extent ever recorded by satellites and about half the average minimum from 1981 to 2010 (from NASA, public domain)

may drift further than thicker sea ice, but the wider distribution is not indicative of greater ice volume. The thinner slabs of sea ice are more easily distributed by wind and currents, may cover a larger area, but account for less volume and the Arctic is rapidly losing sea ice volume.

In late December 2013 and early January 2014, a research vessel sent to the Antarctic to study the effects of global warming became stuck in sea ice. A second ship was sent to rescue the first and also became stuck. Climate change deniers thought this, along with a prolonged cold spell in the U.S., indicated proof that the Earth was not warming. Observations from satellites and ships indicated that the ships were stuck in a massive amount of sea ice broken off from an ice shelf or glacier and had little relationship to global warming or cooling.

## *2.17.3 Ice Shelves*

Ice shelves are platforms of ice that form where ice sheets and glaciers move into the ocean. Ice shelves exist mostly in Antarctica and Greenland, as well as in the Arctic near Canada and Alaska. They are mainly an extension of glacial ice on land, so their melting raises sea level. In the colder months, sea ice may form in the vicinity of the ice shelf. When an ice shelf melts, it allows glacial ice to move seaward to replace it and this additional ice raises sea level when it disintegrates. This is happening in Greenland and Antarctica.

Research by glaciologists suggests that glaciers behind ice shelves may accelerate by as much as five times following a rapid ice shelf retreat or collapse. In recent years, ice shelves on the Antarctic Peninsula and along the northern coast of Canada have experienced rapid disintegration. In March 2008, the Wilkins Ice Shelf in Antarctica retreated by more than 400 square kilometers (160 square miles). Later that summer, several ice shelves along Ellesmere Island in Northern Canada broke up in a matter of days (NSIDC 2013).

In contrast, the collapses in previous years happened over a period of weeks, leaving chunky ice and small icebergs. The remaining ice shelves retreated by as much as 90 % and several have experienced repeated collapses (NSIDC 2013).

The northern section of the Larsen B ice shelf, a large floating ice mass on the eastern side of the Antarctic Peninsula, shattered and separated from the continent (in 2002) in the largest single event in a 30-year series of ice shelf retreats in the peninsula.

### 2.17.4 Icebergs

An iceberg is a body of frozen fresh water that has broken off from a glacier, a larger iceberg, or ice shelf and floats freely in the ocean or lake. Because of density differences, only about 1/10 of an iceberg floats above water, and as a result, icebergs are hazardous to vessels at sea. The North Atlantic and the cold waters surrounding Antarctica are home to most of the icebergs on Earth.

The process that forms icebergs is called calving. Icebergs form when calving takes place along glacial fronts, from ice shelves, and from larger icebergs.

### 2.17.5 Current and Recent Research

Those agencies cited above in the section on glaciers; NASA, NOAA, universities, and geological surveys monitor and map sea ice. The University of Washington and the University of Otago, Dunedin, New Zealand have active research programs concerning sea ice.

Icebergs are monitored worldwide by the U.S. National Ice Center (NIC). NIC produces analyses and forecasts of Arctic, Antarctic, Great Lakes, and Chesapeake Bay ice conditions. NIC is the only organization that names and tracks all Antarctic Icebergs.

## *2.17.6 Permafrost*

Permafrost is ground that is frozen from 1 year to the next, often defined as frozen for two consecutive years. It is found in high latitudes and high altitudes.

Permafrost often contains information about past climate change. Some permafrost has been frozen for tens or hundreds of thousands of years. This old permafrost may contain material from plants or animals that used to live in the area. These remains give hints of what the climate was like when they lived. Scientists also study chemistry of the ice making up permafrost and can learn what the atmosphere was like when the permafrost formed.

Changes in frozen soils are a strong indication of climate change. In remote, cold regions, where people have not taken many temperature measurements in the past, frozen ground is an especially important record. Permafrost regions occupy approximately 22.79 million square kilometers (about 24 % of the exposed land surface) of the Northern Hemisphere and permafrost underlies the ice caps of Greenland and Antarctica, and individual glaciers throughout the world.

Despite its name, permafrost is unstable and not permanent. It is often covered by an "active layer" that regularly melts during the warmer season. Although permafrost can be thousands of years old, it is sometimes newly formed or about to thaw, and it often exists close to its melting point. Its upper few centimeters are referred to as the "active zone" that melts during warmer periods and refreezes during the next cold period.

Permafrost's "active zone" wreaks havoc on construction in areas where it occurs. Roads and buildings are destroyed as the permafrost melts and landslides occur in mountainous regions. Field observations indicate that permafrost warmed by up to 6 °C during the 20th century. Observations on Svalbard detected extreme permafrost warming during the winter-spring 2005–2006. The thaw apparently resulted from a temperature anomaly where January and April temperatures reached more than 12 °C above the 1961–1990 average.

Observations in Alaska found permafrost warming at most sites north of the Brooks Range from the Chukchi Sea to the border with Canada coincident with statewide air-temperature warming beginning in 1976. The warming occurred primarily in the winter, with little summertime change. This melting permafrost is greatly increasing runoff of fresh water to the Arctic Ocean.

The Arctic contains nearly one-third of the Earth's stored soil carbon. If the high northern latitudes continue to have a significant temperature increase, the regional soils will begin to release carbon into the atmosphere, which will lead to higher temperatures, fueling the cycle of carbon release and temperature rise. Earth may have already reached a tipping point concerning permafrost as there are numerous reports of methane bubbling to the surface in Arctic area lakes.

Northern latitudes are warming at least twice (2–4 times) as fast as other parts of the world. Included with permafrost in these latitudes is a vast store of carbon from plants and animals that lived in these areas before formation of the permafrost, or lived in the active layer in warmer months and were encapsulated during refreezing in colder months as the permafrost was being built up. As warming continues and permafrost breaks down, carbon is released largely in the form of methane ($CH_4$), a greenhouse gas about 84 times more potent over a 20 year period than carbon dioxide (IPCC 2013).

Some researchers have calculated that there are about 1.7 trillion tons of carbon in soils of the northern regions, about 88 % of it locked in permafrost. That is about two and a half times the amount of carbon already in the atmosphere.

Wildfires are increasing across much of the north, and early research suggests that extensive burning could lead to a more rapid thaw of permafrost and release of methane. Fires occur in the warmer months when vegetation takes root in the upper layers. The wildfires hasten melting of permafrost and increase releases of carbon. Wildfires are often set off by lightning and burn vast areas of tundra.

Parts of the tundra are becoming greener each year as more southerly plants migrate northward in the Northern Hemisphere and southward in the Southern Hemisphere.

### 2.17.7  Current and Recent Research

Active research teams that are currently studying the changing climate of northern latitudes are from various universities and government agencies including the University of Alaska, University of Florida (a surprise), U.S. Department of Energy, NOAA, NASA, the International Arctic Research Center (IARC), the U.S. Arctic Research Commission, the U.S. National Science Foundation (NSF), The Institute of Arctic and Alpine Research at the University of Colorado, Boulder, University of Illinois, NSIDC, Arctic Research in Finland (multiple entities), Developing Arctic Modeling and Observing Capabilities for Long-term Environmental Studies (DAMOCLES), European Space Agency (ESA), and the Russian Academy of Sciences. This is not intended to be a complete list but a guide to those seeking additional information on Arctic research.

## 2.18  Rising Sea Level

Sea level has been rising worldwide since the Last Glacial Maximum (LGM) about 20,000–18,000 years ago. As Earth temperatures are rising relatively rapidly, sea level is rising relatively rapidly (Fig. 2.8).

To measure sea level, oceanographers at NASA's Jet Propulsion Laboratory (JPL) rely on satellite measurements of sea surface height (which increases as temperature increases) taken by TOPEX/Poseidon and later by Jason-1 satellites. Complementing the Jason-1 satellite data were temperature and salinity

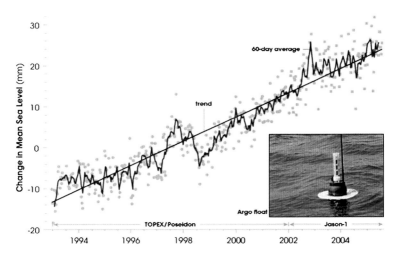

**Fig. 2.8** Relying on data from satellites and Argo floats (mechanical devices drifting in the ocean, one of which is shown *above*), a group of oceanographers announced in June 2006 that sea level rose, on average, 3 mm (0.1 in.) per year between 1993 and 2005. This graph shows the increase in mean sea level, measured in millimeters. Researchers attributed about half of that increase to melting ice and the other half to thermal expansion as the ocean absorbs excess energy (from NASA, public domain)

measurements from the Argo float program. By using measurements from a variety of sources, oceanographers can form a clearer picture of the ocean's behavior in different parts of the world.

Another tool useful in the study of sea level is NASA's Gravity Recovery and Climate Experiment (GRACE). GRACE consists of twin satellites that precisely measures surface height not only of the world's ocean, but also the giant bodies of ice that feed it. If ice mass height drops and ocean level rises, GRACE can measure both changes simultaneously. GRACE observations determined that from 2002 to 2005, Antarctic ice lost enough mass to raise global sea level by 1.5 mm (0.05 in.); Earth's ice sheets are melting at the alarming rate of 300 billion tons per year.

In 2012, lower than average sea levels around the west coast of North America were linked to the cool temperatures associated with the negative phase of the Pacific Decadal Oscillation (PDO) (see Fig. 2.9). The sharp divide in the North Atlantic between areas of above-average and below-average sea level are evidence that the warm waters of the Gulf Stream Current reached farther north than usual.

The higher-than-normal sea levels in the western tropical Pacific and Indian Ocean are an imprint left by the La Niña event that continued from 2011 into the first part of 2012. During La Niña, the trade winds along the equator blow more strongly than normal, pushing warm tropical waters westward across the Pacific. The warm waters "pile up" in the western part of the tropical Pacific, spilling over into the Indian Ocean, making sea levels there higher than average (NOAA 2013).

January-December 2012

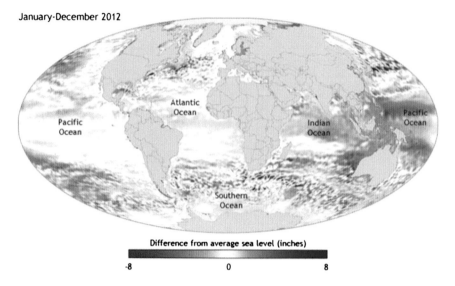

**Difference from average sea level (inches)**

-8                          0                          8

**Fig. 2.9** Sea level in 2012 compared to the 1993–2012 average based on AVISO satellite data. Map adapted from Fig. 3.27a in the 2012 BAMS State of the climate report (from NOAA, public domain)

## 2.19 The Paleocene-Eocene Thermal Maximum

The Paleocene-Eocene Thermal Maximum (PETM) is given a good deal of attention here because the PETM is the definitive geologic analog to Earth's present climatic situation; a time of rapidly increasing global temperature, a rapid rise in sea level, and a rapid increase of carbon dioxide in the atmosphere.

The PETM extreme global warm-up happened 56 million years ago, when Pangaea was splitting into separate continents. It is suspected that huge amounts of carbon were released into the atmosphere and ocean in the form of carbon dioxide and methane. The globe warmed 5–9 °C (9–16 °F). Most ecosystems were able to adapt. But the rate of warming during the PETM pales in comparison to what we're now experiencing. Today, global temperature could be warming at a rate that is too fast for ecosystems to adapt; too fast for land and sea life to migrate. In fact, the current warming may be taking place at a pace not like anything that the Earth has experienced in millions of years. The PETM warming was a roughly 200,000-year long event.

Global temperatures rose by about 6 °C (11 °F) and humans are on track to possibly reach this number. It is for this reason that the PETM is receiving so much attention by global warming researchers. It is now widely accepted that the PETM represents a "case study" for global warming and massive carbon input to Earth's atmosphere.

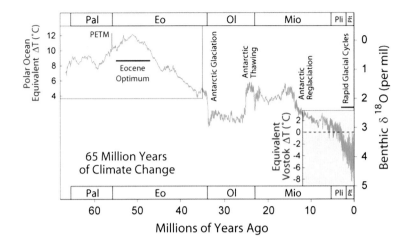

**Fig. 2.10**   Climate change during the last 65 million years as expressed by the oxygen isotope composition of benthic foraminifera. The Paleocene-Eocene Thermal Maximum (*PETM*) is characterized by a brief but prominent negative excursion, attributed to rapid warming. Note that the excursion is understated in this graph due to the smoothing of data. This figure was prepared by Robert A. Rohde from published and publicly available data and is incorporated into the Global Warming Art project

There are three naturally occurring isotopes of carbon, with $^{12}$C (carbon-12) and $^{13}$C (carbon-13) being stable, while $^{14}$C (carbon-14) is radioactive, decaying with a half-life of about 5,730 years. The isotope carbon-12 ($^{12}$C) forms 98.93 % of the carbon on Earth, while carbon-13 ($^{13}$C) forms the remaining 1.07 %. The concentration of $^{12}$C is further increased in biological materials because biochemical reactions discriminate against $^{13}$C (Fig. 2.10).

Although $^{13}$C only constitutes 1.07 % of carbon on Earth, it is still detectable by geochemical analysis. During the PETM there was a distinct reduction in the $^{13}$C isotope ($\delta^{13}$C; change in $^{13}$C). At the start of the PETM there was a massive input of $^{13}$C-depleted carbon that entered the hydrosphere or atmosphere. Estimates of the relatively sudden addition range from about 2,500 to over 6,800 gigatons of $^{13}$C depleted carbon.

Evidence for a massive addition of $^{13}$C-depleted carbon at the onset of the PETM comes from two distinct observations. First, a prominent negative excursion in the carbon isotope composition ($\delta^{13}$C) of carbon-bearing phases characterizes the PETM in numerous widespread locations from a range of environments. Second, carbonate dissolution marks the PETM in sediment cores from the ocean basins.

The Paleocene-Eocene Thermal Maximum (PETM) is the most recent event that we can compare to today's warming. Global temperatures rose at least 5 °C (9 °F), and the PETM warmth lasted 200,000 years before the Earth system was able to remove the extra $CO_2$ from the atmosphere. The resulting impact on Earth's climate was so severe that a new geological epoch was born, the Eocene.

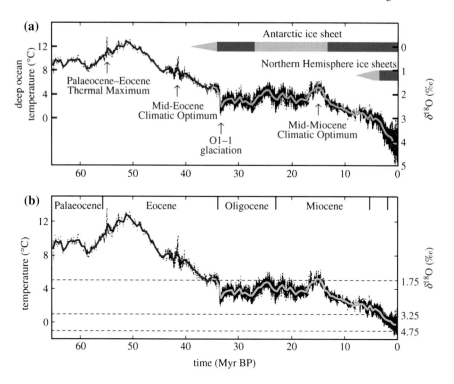

**Fig. 2.11** **a** Global deep ocean $\delta^{18}O$ from Zachos et al. and **b** estimated deep ocean temperature. *Black data points* are five-point running means of the original temporal resolution; *red (left)* and *blue (right) curves* have a 500 kiloyear resolution (Hansen et al. 2013)

## 2.19.1 Global Warming 56 Million Years Ago

After years of research, the PETM is now thought to have been caused by greenhouse gas emissions, similar to how Earth is warming today. About 56 million years ago, at the end of the Paleocene Epoch (Fig. 2.11; see Appendix A), the supercontinent Pangaea was in the final stages of breaking apart into the continents as we know them today. As the land masses split apart (land masses pulling apart are "stretched" causing a build-up of tensile forces), volcanoes erupted and molten rock seeped toward the Earth's surface, literally baking carbon-rich sediments and releasing carbon dioxide into the air. During this time, global atmospheric temperature probably increased by several degrees.

The initial increase in temperature triggered events that led to more greenhouse gas emissions and more warming (i.e., positive feedbacks). Climate scientists generally agree that the feedback with the most impact on the atmospheric temperature increase was the melting of methane hydrates in the ocean basins. As the atmosphere warmed the ocean surface, currents (probably not unlike the thermohaline circulation we know today) would have funneled the warm water to the ocean floor (due to increased density), where it melted the frozen methane hydrates (also referred to as methane clathrates), releasing the potent greenhouse gas (methane) into the ocean and eventually bubbling up to the atmosphere.

## 2.19.2  PETM Versus Modern Greenhouse Gas Emissions

Temperature rose steadily in the PETM due to the slow release of greenhouse gases (around 2 billion tons per year). Today, fossil fuel burning is leading to 30+ billion tons of carbon released into the atmosphere every year, driving global temperature up at an incredible rate and a rate unknown certainly in human history.

Many of the other climate feedbacks that we either already observe today or expect to experience probably took place during the PETM warming, as well.

## 2.19.3  PETM Warming Versus Current Warming

During the PETM, around 5 billion tons of $CO_2$ was released into the atmosphere per year. The Earth warmed around 6 °C (11 °F) over 200,000 years; some estimates are that the warming was more like 9 °C (16 °F). Using the low end of that estimated range, the globe warmed around 0.025 °C per year for 100 years. Today, the globe is warming at least ten times as fast, anywhere from 1 to 4 °C every 100 years. In 2010, our fossil fuel burning released 35 billion tons of $CO_2$ into the atmosphere. By comparison, volcanoes release only about 0.2 billion tons of $CO_2$ per year. How fast carbon enters the atmosphere translates to how fast temperature increases, and the environmental and societal consequences of warming at such a speed could be devastating to humankind.

Some of the things that we fear might happen in the not too distant future such as reversed ocean circulation, ocean acidified, permafrost melted, peatlands and forests burned in wildfires happened during the PETM.

## 2.19.4  PETM Ocean Circulation

There is evidence that during the PETM, large-scale ocean circulation may have reversed, which would have led to enhanced warming. Ocean circulations are largely due to atmospheric circulation, temperature and salinity (salt concentration), and warming of ocean water at high latitudes would have acted to at least slow, if not totally reverse, the "global conveyor belt" (the thermohaline circulation that occurs in the modern ocean).

## 2.19.5  PETM Sea Level Rise

Since the PETM occurred in an already warm climate (that sets the PETM apart from modern warming), there was very little sea ice and glacial cover to melt, so sea level did not change dramatically. However, there is plenty of ice to melt today on the planet and it is already melting.

## 2.19.6  PETM Permafrost and Methane Hydrates

Many scientists agree that a major contributor to the PETM warming was the melting of methane hydrates on the seafloor and permafrost at high latitudes. Both of these store immense amounts of carbon deficient in carbon-13 and today constitute a tipping point for the climate. Once hydrates and permafrost begin to melt, the process will be irreversible (permafrost has already started to melt). The reservoirs of methane hydrate stored in marine sediments (500–10,000 billion tons of carbon) and in permafrost (7.5–400 billion tons) are being constantly monitored.

## 2.19.7  PETM Ocean Acidification

The most disruptive impact during the PETM was likely the exceptional ocean acidification. The ocean naturally absorbs carbon dioxide from the atmosphere, and also from the sea floor (in the form of calcium carbonate dissolution). When excess carbon dioxide enters the atmosphere, the ocean attempts to balance the system by absorbing more of it. Numerous studies have shown that this was the case during the PETM.

## 2.20  Species Extinction

Climate change/global warming is changing the environment of all living things on Earth. Those organisms with narrow limits of tolerance will be the first to go extinct and many already have. Extinction is permanent unless DNA is somehow preserved.

Specific examples of species at risk for physiological reasons include mountain species such as pikas (small bunny-like lagomorphs that only live in cold climates) and endemic Hawaiian Mauna Kea silverswords, which are restricted to cool temperatures at high altitudes. Species like polar bears are at risk because they depend on sea ice to facilitate their hunting of seals and Arctic sea ice conditions are changing rapidly, as we have seen. Other species are prone to extinction as changing climate causes their habitats to alter such that growth, development, or reproduction of individuals are inhibited.

The distinct risks of climate change exacerbate other widely recognized and severe extinction pressures, especially habitat destruction, competition from invasive species, and unsustainable exploitation of species for economic gain, which have already elevated extinction rates to many times above background rates. If unchecked, habitat destruction, fragmentation, and over-exploitation, even without climate change, could result in a mass extinction within the next few centuries equivalent in magnitude to the one that wiped out the dinosaurs. With the ongoing pressures of climate change, comparable levels of extinction conceivably could occur before the year 2100; indeed, some models show a crash of coral reefs from climate change alone as early as 2060 under certain scenarios.

Loss of a species is permanent and irreversible, and has both economic impacts and ethical implications. The economic impacts derive from loss of ecosystem services, revenue, and jobs, for example in the fishing, forestry, and ecotourism industries. Ethical implications include the permanent loss of irreplaceable species and ecosystems as the current generation's legacy to the next generation.

Earth is currently experiencing its sixth mass extinction of plants and animals, the sixth wave of extinctions in the past half-billion years. The planet is currently experiencing the worst of species die-offs since the loss of the dinosaurs 65 million years ago and very little is presently being done about it.

Because the rate of change in our biosphere is increasing, and because every species' extinction potentially leads to the extinction of others bound to that species in a complex ecological web, numbers of extinctions are likely to snowball in the coming decades as ecosystems unravel due to global warming, either directly or indirectly.

Dr. Terry Root, Professor and Senior Fellow at the Stanford Woods Institute for the Environment, is blunt about the evidence she's collected in her studies of the effect of temperature on wildlife. Root says we're en route to a mass extinction in which as many as half the world's species will die out, and she sees no way off that path.

The IPCC tells us that this century could see the extinction of 50 % of the species now living on Earth. It is not a pleasant scenario.

## 2.21  Deforestation

Cutting down trees has been characteristic of areas humans have invaded throughout the history of the species. Humans have always needed room for their various activities and trees just "got in the way." This was certainly the case when humans first left East Africa and travelled through the Middle-East and into forest-covered Europe and Asia, then into North and South America. They cut down trees to expand their settlements, their hunting and gathering areas, and later for agriculture.

The Amazon rain forest is one of the most diverse places on Earth, home to unknown thousands of species, including thousands of trees. Until just recently, scientists didn't know which trees dominated the forest and where different tree species were located. A new study published in the Oct. 17, 2013, issue of the journal *Science* examined an area of the Amazon and found that only a few dozen tree species are very common, or "hyperdominant."

With help from Google Earth Engine, the World Resources Institute launched Global Forest Watch, an online forest monitoring and alert system that allows individual computer users to watch forests around the world change in an almost real-time stream of imagery. Additional information can be found at the following website: http://www.wri.org/events/gfwlaunch.

## 2.22  Ecosystem Impacts

Some of the impacts of global warming already imposed on ecosystems today are such things as ecosystem processes that control expansion, perpetuity, and decomposition. There have been shifts in the geographic ranges of aquatic species, some cold water species have had their ranges restricted while warm water species have seen their ranges expand. Animals and plants are migrating to higher elevations and latitudes and leaving behind those that can't migrate.

Seasons of the year are changing with summers getting longer and winters shorter and this is disrupting breeding seasons and migratory patterns of birds. The season changes are also causing insect pests' lifespans, like the pine bark beetle in the Rocky Mountains of western North America, to continue so that one generation overlaps another thereby rapidly increasing the population of hungry beetles.

Forest fires are wreaking havoc throughout western North America, Russia, Australia, and elsewhere and thereby destroying niches and habitats in their paths. Disease pathogens are spreading from the equatorial regions to higher latitudes.

Deserts and dry lands are likely to become hotter and drier, feeding a cycle of invasive plants, fire, and erosion. Coastal and near-shore ecosystems are already under multiple stresses from rising sea level and pollution. Climate change and ocean acidification will worsen these stresses.

Arctic sea ice ecosystems are already being adversely affected by the loss of summer sea ice and further changes are expected as the Arctic continues to warm about twice as fast as the rest of the world.

The habitats of some mountain species, such as the pika, and cold water fish, such as salmon and trout, are very likely to contract in response to global warming. Some of the benefits ecosystems provide to society will be threatened by climate change, while others will be enhanced. There are already migration of various species of land mammals poleward or to higher elevations.

Extinction of some benthic foraminifera (bottom-dwelling single-celled organisms with shells or tests). Coral bleaching (dead corals are bleached white) due to ocean acidification has been observed in large numbers of coral reef ecosystems.

Famine and malnutrition due to drought have been found in species in threatened ecosystems. Species endangerment (e.g., polar bears, marine turtles, North Atlantic whales, giant pandas, orangutans, elephants) is increasing. There has been increased mortality in organisms from extreme weather and malnutrition, increase in disease vectors, decrease in agricultural yield, mass wildlife migration and extinction, total societal collapse in many ecosystems.

## 2.23  Research Programs

At the national level in the United States, the U.S. Global Change Research Program coordinates the world's most extensive research effort on climate change. In addition, NASA, NOAA, the U.S. Environmental Protection Agency (EPA), the U.S. Geological Survey (USGS) and other federal and state agencies are actively engaging the private sector. Many university and private scientists also study climate change. In the U.S., universities with well-established programs in climate science are Rutgers University, Pennsylvania State University, University of New Hampshire, University of Chicago, Oregon State University, University of Virginia, University of California-San Diego, University of Washington, Louisiana State University, and Ohio State University. There are also numerous other U.S. universities and colleges that have relatively new programs in climate science.

In the U.K., The Met Office hosts the National Climate Information Centre (U.K. NCIC) which holds national and regional climate information for the United Kingdom. Digitized records for the whole country date back to 1910 and data for the Central England Temperature record dates back to 1654, the world's longest instrumental record.

The U.K.'s University of East Anglia has the Climate Research Unit (CRU), one of the world's leading institutions concerned with the study of natural and anthropogenic climate change. Other U.K. universities with well-established climate science programs are University College London, University of Exeter, University of Leicester, Imperial College London, King's College London, University of Cambridge, and the University of Oxford.

In Australia, University of Melbourne, University of Queensland, University of New South Wales, Macquarie University, Sydney, Monash University, University of Western Australia, and the Australian National University, Canberra.

Other countries have similar climate research programs.

## 2.24  Status of Climate Change Science—End of February 2014

The status of climate change science at the end of February 2014 is, in one word, strong. The subject has attracted some of the best scientific minds in the world. The problems faced by those who study climate are not with the science, they are with the fossil fuel industry and the doubters controlled or influenced by them.

The current status of climate change science at the time of this writing (February 2014) is that it is progressing well with many research projects underway to answer many of the questions that remain unresolved. A great number of modeling experiments are being completed. Prominent climate scientists are an impressive group and listing their names would not serve any purpose. They know who they are.

Climate scientists have been warning of the dangers of global warming especially since the 1988 appearance of James Hansen before Congress. There were warnings before 1988 but they were largely ignored. Unfortunately, most of the warnings today are also largely ignored. Most politicians in particular don't seem to understand the science, especially those with conservative ideologies even though the scientific evidence of global warming happening now is overwhelming. Global warming does not favor any one politic or persuasion. It does not care whether you are conservative or liberal, white, black, green or brown. It is affecting everyone on Earth to some degree; and climate scientists are telling the world to get ready for the next degree because the warming will only get worse with time.

What is it going to take to get the attention of climate change/global warming deniers in the U.S. Congress? This writer's efforts are directed at getting them out of office by using the ballot box as soon as possible; and I hope others' efforts are the same.

The 19th Conference of the Parties of the UN Framework.

Convention on Climate Change (COP 19), held in November 2013 in Warsaw, Poland, was warned by professor of climatology Mark Maslin (University College London): "We are already planning for a 4 °C world because that is where we are heading. I do not know of any scientists who do not believe that."

The following is a bulleted attempt to bring the reader up to date on climate change science from January 2013 (the publication of Farmer and Cook's Climate Change Science: A Modern Synthesis, January 12 2013) through February 2014. The bulleted items below are not in chronological order or in order of importance, but they are all significant to climate change.

Some of these items, following sometimes disturbing results of recent research (January 2013–February 2014), have been shared with me by Professor Guy McPherson, Professor Emeritus of Natural Resources and Ecology and Evolutionary Biology, University of Arizona, Tucson. Professor McPherson has kindly allowed me editorial license which I have used sparingly. Any errors of omission or commission are the property of this writer and not Professor McPherson.

- Nearly 500 climate-related laws have been enacted in 66 countries that together are responsible for nearly 90 % of the world's heat-trapping emissions, according to an international survey of climate legislation released February 27 2014 by the London-based Global Legislators Organisation (GLOBE) and the Grantham Institute at the London School of Economics. (From *Nature*, "National policies advance as climate summit approaches," News release February 27 2014).
- As reported in the journal *Global and Planetary Change* in April 2013, almost every molecule of atmospheric carbon dioxide ($CO_2$) added since 1980 has come from human emissions. A few molecules from volcanic eruptions may also be present in the atmosphere.
- Carbon dioxide concentration in the atmosphere has now topped 400 ppm. On May 9, 2013 the daily mean concentration of carbon dioxide in the atmosphere of Mauna Loa, Hawaii, surpassed 400 parts per million (ppm) for the first time since measurements began in 1958. Independent measurements made by both NOAA and the Scripps Institution of Oceanography were approaching this level during the

previous week. It marks an important milestone because Mauna Loa, as the oldest continuous carbon dioxide ($CO_2$) measurement station in the world, is the primary global benchmark site for monitoring the increase of this potent heat-trapping gas.

- Methane hydrates are bubbling out of the Arctic Ocean (*Science*, March 2010). According to NASA's CARVE project, these plumes were up to 150 km across as of mid-July 2013. In early November 2013, methane levels well in excess of 2,600 ppb were recorded at multiple altitudes in the Arctic. Later that same month, Shakhova and colleagues (see Shakhova below) published a paper in *Nature Geoscience* suggesting "significant quantities of methane are escaping the East Siberian Shelf" and indicating that a 50-billion-tonne "burst" of methane could warm Earth by 1.3 °C. Such a burst of methane is "highly possible at any time." By 15 December 2013, methane bubbling up from the seafloor of the Arctic Ocean had sufficient force to prevent sea ice from forming in the area. The "methane bomb hypothesis" is not dead yet. (Dr. N. Shakhova leads the Russia-U.S. Methane Study at the International Arctic Research Center, at the University Alaska Fairbanks and the Pacific Oceanological Institute, Far Eastern Branch of Russian Academy of Sciences).
- Methane seeps are appearing in numerous locations off the east coast of the United States, leading to rapid destabilization of methane hydrates (according to the 25 October 2013 issue of *Nature*).
- On land, anthropogenic emissions of methane in the United States have been severely underestimated by the Environmental Protection Agency, according to a paper in the 25 November 2013 issue of *Proceedings of the National Academy of Sciences*. This figure is 1,100 ppb higher than pre-industrial peak levels. Methane release tracks closely with temperature rise throughout Earth history—specifically, Arctic methane release and rapid global temperature rise are interlinked—including a temperature rise up to about 1 °C per year over a decade, according to data from ice cores.
- Methane is being released from the Antarctic, also (*Nature*, August 2012). According to a paper in the 24 July 2013 issue of *Scientific Reports*, melt rate in the Antarctic has caught up to the Arctic and the West Antarctic Ice Sheet is losing over 150 cubic kilometers of ice each year according to the European Space Agency's CryoSat observations published 11 December 2013, and Antarctica's crumbling Larsen B Ice Shelf is poised to finish its collapse, according to a glaciologist at the *National Snow and Ice Data Center* (*NSIDC*) at the annual meeting of the American Geophysical Union (AGU) December 2013.
- The Intergovernmental Panel on Climate Change (*IPCC*) admits global warming is irreversible without geoengineering in a report released 27 September 2013.
- As pointed out in the 5 December 2013 issue of *Earth System Dynamics*, known strategies for geoengineering are unlikely to succeed.
- As published in the 1 August 2013 issue of *Science*, in the near term Earth's climate will change orders of magnitude faster than at any time during the last 65 million years, since the extinction of the dinosaurs.
- Without the large and growing number of self-reinforcing feedback loops humans have triggered recently, the 5 °C rise in global average temperature 55 million years ago during a span of just a few years, and it looks like we should be addressing the 6–10 °C of warming we are working toward.

- *The New Yorker* posits a relevant question on 5 November 2013: "Is It Too Late to Prepare for Climate Change?"
- The *Geological Society of London* points out on 10 December 2013 that Earth's climate "could" be twice as sensitive to atmospheric carbon as previously believed.
- According to Yvo de Boer, who was executive secretary of the *United Nations Framework Convention on Climate Change* in 2009, when attempts to reach a deal at a summit in Copenhagen crumbled with a rift between industrialized and developing nations, "the only way that a 2015 agreement can achieve a 2-degree goal is to shut down the whole global economy."
- Drought in the Amazon triggered the release of more carbon than the United States in 2010 (*Science*, February 2011). In addition, ongoing deforestation in the region is driving declines in precipitation at a rate much faster than long thought, as reported in the 19 July 2013 issue of *Geophysical Research Letters*.
- Russian forest and bog fires are growing (NASA, August 2012), a phenomenon consequently apparent throughout the northern hemisphere (*Nature Communications*, July 2013).
- A paper in the 22 July 2013 issue of the *Proceedings of the National Academy of Sciences* indicates boreal forests are burning at a rate exceeding that of the last 10,000 years.
- Exposure to sunlight increases bacterial conversion of exposed soil carbon, thus accelerating thawing of the permafrost and the release of methane (*Proceedings of the National Academy of Sciences*, February 2013).
- Summer ice melt in Antarctica is at its highest level in a thousand years. Summer ice in the Antarctic is melting 10 times quicker than it was 600 years ago, with the most rapid melt occurring in the last 50 years (*Nature Geoscience*, April 2013).
- Although scientists have long expressed concern about the instability of the West Antarctic Ice Sheet (WAIS), a research paper published in the 28 August 2013 of *Nature* indicates the East Antarctic Ice Sheet (EAIS) has undergone rapid changes in the past five decades. The latter is the world's largest ice sheet and was previously thought to be at little risk from climate change. But it has undergone rapid changes in the past five decades, signaling a potential threat to global sea levels. The EAIS holds enough water to raise sea levels more than 50 m.
- Increased temperature and aridity in the southwestern interior of North America facilitates movement of dust from low-elevation deserts to high-elevation snow-pack, thus accelerating snowmelt, as reported in the 17 May 2013 issue of *Hydrology and Earth System Sciences*.
- Floods in Canada are sending pulses of silty water out through the Mackenzie Delta and into the Beaufort Sea, thus painting brown a wide section of the Arctic Ocean near the Mackenzie Delta (NASA, June 2013).
- Surface meltwater draining through moulins and cracks in an ice sheet can warm the sheet from the inside, softening the ice and letting it flow faster, according to a study accepted for publication in the *Journal of Geophysical Research: Earth Surface* (July 2013). It appears a Heinrich Event has been triggered in Greenland. Consider the description of such an event as provided by Robert Scribbler on 8 August 2013: "In a Heinrich Event, the melt forces eventually reach a tipping point. The warmer water has greatly softened the ice sheet. Floods of water flow

out beneath the ice. Ice ponds grow into great lakes that may spill out both over top of the ice and underneath it. Large ice dams may or may not start to form. All through this time ice motion and melt is accelerating. Finally, a major tipping point is reached and in a single large event or ongoing series of such events, a massive surge of water and ice flush outward as the ice sheet enters an entirely chaotic state. Tsunamis of melt water rush out bearing their vast flotillas of icebergs, greatly contributing to sea level rise. And that's when the weather really starts to get nasty. In the case of Greenland, the firing line for such events is the entire North Atlantic and, ultimately the Northern Hemisphere."

- Breakdown of the thermohaline conveyor belt is happening in the Antarctic as well as the Arctic, thus leading to melting of Antarctic permafrost (*Scientific Reports*, July 2013).
- Ocean acidification leads to release of less dimethyl sulfide (DMS) by plankton. DMS shields Earth from radiation. (*Nature Climate Change*, online 25 August 2013). Plankton forms the base of the marine food web, and are on the verge of disappearing completely, according to a paper in the 17 October 2013 issue of *Global Change Biology.*
- Rising ocean temperatures will upset natural cycles of carbon dioxide, nitrogen and phosphorus, hence reducing plankton (*Nature Climate Change*, September 2013). Plankton forms an essential base of the food chain.
- Small ponds in the Canadian Arctic are releasing far more methane than expected based on their aerial cover (*PLoS ONE*, November 2013).
- Arctic ice is growing darker, hence less reflective (*Nature Climate Change*, August 2013) and thus increasing Arctic warming.
- The *New York Times* reports hotter, drier conditions leading to huge fires in western North America as the "new normal" in their 1 July 2013 issue.
- Loss of Arctic sea ice is reducing the temperature gradient between the poles and the equator, thus causing the jet stream to slow and meander. One result is the creation of weather blocks such as the recent very high temperatures in southern Alaska. As a result, boreal peat dries and catches fire like a coal seam. The resulting soot enters the atmosphere to fall again, coating the ice surface elsewhere, thus reducing albedo and hastening the melting of ice. Each of these individual phenomena has been recently reported (July 2013).
- Plankton forms the base of the marine food chain and is on the verge of disappearing completely, according to a paper in the 17 October 2013 issue of *Global Change Biology.*
- Sea-level rise causes slope collapse, tsunamis, and release of methane, as reported in the September 2013 issue of *Geology.* In eastern Siberia, the speed of coastal erosion has nearly doubled during the last four decades as the permafrost melts.
- Meanwhile, Arctic drilling was fast-tracked by the Obama administration during the summer of 2012; but Shell Oil has withdrawn their plans to begin drilling in the summer of 2014.
- The IPCC, on 30 September 2013, released a final draft of the first volume of their latest Assessment Report (AR5), *The Physical Science Basis*, and the first of several volumes making up the AR5. The remaining volumes will be released during 2014.

- The IPCC changes made from the last report, AR4 released in 2007, are many but two stand out: (1) a revision in their estimate of sea level rise in a worst case, to 1 m by 2100; and (2) a change in climate sensitivity at the low end of the range to 1.5 °C from 2 °C.
- There was great concern throughout 2013 into 2014 about the severe drought in California and the southwestern U.S., and the link to global warming.
- Australia reported its hottest year (2013–2014) in their history of record keeping.
- The British Isles have been ravished (2013–2014) by intense storms causing widespread flooding in the Thames River Valley and elsewhere.
- There was much talk about the "polar vortex" affecting weather in North America during the winter of 2013–2014. One conservative radio commentator stated that liberals had invented the "polar vortex" to explain global warming. "Polar vortex" has been in meteorology books since early 20th century. Climate scientists have long predicted that the Jet Stream would develop large meanders as the result of warming in the Arctic and the lessening of Arctic sea ice. A large meander system of the Jet Stream developed over the U.S. during the winter of 2013–2014 causing record high temperatures in Alaska and record lows and snowfalls in the mid-western and north-eastern U.S. As the large meander dips southward, cold air is found as far south as the Gulf of Mexico and warm air as far north as Alaska.
- Two-thousand-thirteen (2013) was not the hottest year on record, though it ranked in the top 10 in most datasets: (4th in NOAA/NCDC's record, 7th in NASA/GISS, 8th in the UK's Hadley Centre HadCRUT4, and 5th in the new Cowtan and Way record).
- Warming of the lower atmosphere slowed which raised the question: Where was the extra heat going? It prompted deniers to again claim that global warming stopped in 1998. Climate scientists were able to show that the deep oceans continued to warm at an accelerated rate and that there had been previous episodes when the surface temperature leveled out; Global warming continued and even accelerated during the apparent slowdown with the deep ocean waters heating faster than ever before.
- New research by Cowtan and Way, *Quarterly Journal of the U.K.'s Royal Meteorological Society*, November 2013, creates a new temperature dataset by the statistical method known as Kriging for poorly represented areas in the HadCRUT4 dataset. By this statistical method, fairly accurate temperature data were acquired from previously unmeasured Arctic, Antarctic, and central Africa areas, refuting recent thinking on a slow-down in warming, and increasing estimates of rising temperatures globally. Cowtan and Way realized that although satellites might not give good estimates of surface temperature directly, they provide much more accurate estimates of how temperature varied from place to place. It turns out that if the satellites estimate that an area 10 miles away is about 2° colder than a current location, something quite similar will be seen in the data from weather stations 10 miles away. By using surface measurements for specific locations and using the spatial information from the satellites to fill

in missing regions each month, they discovered that they could create an accurate estimate of temperatures in areas with few or no direct temperature measurements available. The statistical method Kriging was previously applied to land temperatures by the Berkeley Earth group to interpolate the surface measurements based on the satellite field.

- Sea ice extent in 2013 in the Arctic was well below the long-term (1978–2000) average, but nowhere near as low as the record-breaking low of 2012. Antarctic sea ice was unusually high in 2013, pushing overall global sea ice close to the long-term average. Arctic sea ice set a record low in 2012 and the fact that the 2013 sea ice extent was greater than 2012 caused climate change deniers to insist that there was no global warming and it had stopped in 1998, and that the increase in 2013 Arctic sea ice "proved it."

- Sea-level rise has continued in the oceans, with 2013 reversing the slight slow-down that occurred in 2012, and again rising in line with the overall trend over the past decade. Sea-level measurements are now primarily taken with satellite altimeters, which have much greater spatial coverage than tide gauges. IPCC revised their projected sea-level rise by 2,100 up to 3 ft (1 m) in the worst-case scenario (the scenario does not include the collapse of polar ice sheets).

- El Niño and La Niña comprise the periodic cycle in ocean temperatures that drives much year-to-year variability in global temperatures. El Niño years tend to be hotter than normal and La Niña years tend to be colder. In 2013 it was neither an El Niño nor a La Niña year. A moderate El Niño is forecast for the latter part of 2014, with some suggesting that the year could end up setting a new record for the hottest year ever.

- One of the most cited papers in 2013 was entitled "Quantifying the consensus on anthropogenic global warming in the scientific literature" (*Environmental Research Letters* 15 May 2013) by John Cook and colleagues from http://www.s kepticalscience.com. This paper concludes, from a rigorous survey and analysis, that 97 % of climate scientists agree that global warming is occurring and that humans are mainly responsible. This conclusion is in line with and complimentary to previous studies by Oreskes and others.

- Kosaka and Xie (2013) showed that changes in the Pacific Ocean could account for most of the short-term global surface temperature changes in a paper published online 28 August 2013 in *Nature*. Quoting the authors, "Our results show that the current hiatus is part of natural climate variability, tied specifically to La-Niña-like decadal cooling… For the recent decade, the decrease in tropical Pacific sea surface temperature has lowered the global temperature by about 0.15 °C compared to the 1990s."

- On January 16, 2014, the U.S. Senate conducted a hearing on a "Review of the President's Climate Action Plan." Professor Andrew Dessler of Texas A&M University testified that current climate change "is a clear and present danger." In the same hearing, Senator Jeff Sessions (R-AL) quoted Roger Pielke, Jr. (who refers to himself as Dr. but has no earned doctorate; he is not a scientist but is a "political scientist") in the latter's July 2013 testimony before the same committee (Oversight Subcommittee of the U.S. Senate's Committee on

Environment and Public Works) and raised an angry firestorm between Dr. John Holdren, the President's Science Adviser and Mr. Pielke, Jr. Mr. Pielke apparently didn't like being called a liar and accused Dr. Holdren of distorting the science. Unfortunately, Mr. Pielke doesn't understand the science. Dr. Holdren is a preeminent scientist and a former president of the American Association for the Advancement of Science.

- A paper in *Nature Climate Change*, February 2014, by Favier et al., "Retreat of Pine Island Glacier controlled by marine ice-sheet instability" stated that at present the Pine Island Glacier in West Antarctica is thinning and its grounding line has retreated. The work uses three ice-flow models to investigate the stability of the glacier and finds that the grounding line could retreat a further 40 km, which is equivalent to a rise in sea level of 3.5–10 mm over a 20 year period. The new analysis of Antarctica's massive ice sheet suggests the ice is melting much faster now and that the damage is "irreversible." Research shows the massive 68,000-square-mile sheet of ice, believed to be the biggest single contributor to sea-level rise in Antarctica, has begun to shed water at a rate not seen before.

- Balmaseda, Trenberth, and Källén, 2013, published "Distinctive climate signals in reanalysis of global ocean heat content" in *Geophysical Research Letters*, article first published online: 10 May 2013 that examined ocean temperatures with a new analysis showing that in the past decade about 30 % of the heat has been distributed to levels below 700 m, where most previous analyses stop.

- Scientists from University of Toronto's Department of Chemistry discovered a novel chemical lurking in the atmosphere that appears to be a long-lived greenhouse gas (LLGHG). The chemical, perfluorotributylamine (PFTBA) is the most radiatively efficient chemical found to date, breaking all other chemical records for its potential to impact climate. Calculated over a 100-year timeframe, a single molecule of PFTBA has the equivalent climate impact as 7,100 molecules of $CO_2$. Their work as of this writing is unpublished.

- The Fall Meeting of the AGU was held in San Francisco in December 2013 with several sessions devoted to climate science.

- The 19th *Conference of the Parties of the UN Framework Convention on Climate Change (COP 19)* was held in November 2013 in Warsaw, Poland. Little of significance was accomplished. The COP 19 conference closed with a deal finalized on a loss and damage mechanism designed to help developing nations cope with climate change impacts. The deal commits countries to a loss and damage mechanism; the "Warsaw International Mechanism for Loss and Damage." The agreement means that developed nations will be committed to providing expertise and aid to countries who suffer from climate-related impacts.

- Global surface temperatures have slowed over the past 10–15 years. Climate scientists are aware that the positive energy imbalance still exists, so the question of where the heat is going was at first a puzzle. Recently, England et al. published a paper in the journal *Nature Climate Change* entitled "Recent intensification of wind-driven circulation in the Pacific and the ongoing warming hiatus" that explains the hiatus in surface warming as due to internal factors shifting the heat into the oceans. In particular, the rate at which the deep oceans have warmed over the past 10–15 years is unprecedented in the past half century.

- Research led by Masahiro Watanabe of the Japanese Atmosphere and Ocean Research Institute suggests that the hiatus in surface warming is mainly due to more efficient transfer of heat to the deep oceans. Consistent with model simulations led by Gerald Meehl, Watanabe finds that we sometimes expect "hiatus decades" to occur, when surface air temperatures don't warm because more heat is transferred to the deep ocean layers.
- A paper published in *Nature* by Yu Kosaka and Shang-Ping Xie (published online: 28 August 2013) from the Scripps Institution of Oceanography found that accounting for the changes in Pacific Ocean surface temperatures allowed their model to reproduce the slowed global surface warming over the past 10–15 years.
- In the January 2014 issue of *Nature*, the gradual warming of the North and Tropical Atlantic Ocean is seen to be contributing to climate change in Antarctica, a team of New York University scientists has concluded. The findings, which rely on more than three decades of atmospheric data, show new ways in which distant regional conditions are contributing to Antarctic climate change. Xichen Li, a doctoral student in NYU's Courant Institute of Mathematical Sciences and the study's lead author. "Moreover, the study offers further confirmation that warming in one region can have far-reaching effects in another."
- In the mid-1970s, the first available satellite images of Antarctica during the polar winter revealed a huge ice-free region within the ice pack of the Weddell Sea. This ice-free region, or polynya, stayed open for three full winters before it closed. Subsequent research showed that the opening was maintained as relatively warm waters churned upward from kilometers below the ocean's surface and released heat from the ocean's deepest reaches. But the polynya, which was the size of New Zealand, has not reappeared in the nearly 40 years since it closed, and scientists have since come to view it as a naturally rare event. Now, however, a study led by researchers from McGill University (Nature Climate Change, 2014) suggests a new explanation: The 1970s polynya may have been the last gasp of what used to be a more common feature of the Southern Ocean, and which is now suppressed due to the effects of climate change on ocean salinity. The ocean's surface has been steadily getting less salty since the 1950s. This lid of fresh water on top of the ocean prevents mixing with the warm waters underneath. As a result, the deep ocean heat has been unable to get out and melt back the wintertime Antarctic ice pack.
- The Arctic-wide melt season has lengthened 5 days/decade from 1979 to 2013, dominated by later autumn freeze-up within the Kara, Laptev, East Siberian, Chukchi, and Beaufort seas between 6 and 11 days/decade. The timing of melt onset has a large influence on the total amount of solar energy absorbed during summer. The additional heat stored in the upper ocean of approximately 752 MJ m$^{-2}$ during the last decade increases sea surface temperatures by 0.5–1.5 °C and largely explains the observed delays in autumn freeze-up within the Arctic Ocean's adjacent seas. This extra solar energy is equivalent to melting 0.97–1.3 m of ice during the summer (Stroeve et al. 2014 *Geophysical Research Letters*: Article first published online 22 February 2014).
- Google selects Berkeley Earth maps for the Google Maps Gallery launch on February 27 2014.

- The BEST project announced that their new dataset combing land and SST data had been posted.

There are many important papers published and events regarding our changing climate that took place during 2013 and into February 2014 and the above list is but a small sample.

## 2.25  Projections of Future Climate

Figure 2.12 shows observations of (a) Global average surface temperature. (b) Global average sea level, and (c) Northern Hemisphere snow cover from 1850 to 2006.

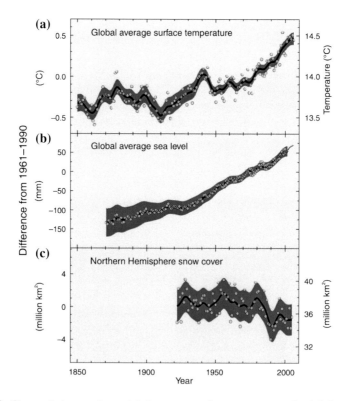

**Fig. 2.12**  Observed changes in **a** global average surface temperature; **b** global average sea level from tide gauge (*blue*) and satellite (*red*) data; and **c** Northern Hemisphere snow cover for March-April. All differences are relative to corresponding averages for the period 1961–1990. Smoothed *curves* represent decadal averaged values while *circles* show yearly values. The *shaded areas* are the uncertainty intervals estimated from a comprehensive analysis of known uncertainties (From IPCC AR4 2007 WG1 synthesis report)

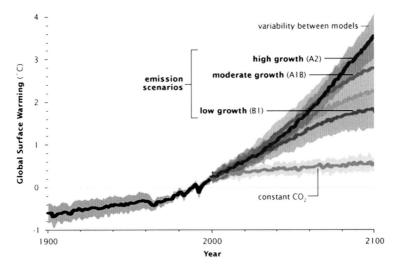

**Fig. 2.13** Model simulations by the intergovernmental panel on climate change estimate that earth will warm between 2 and 6 °C over the next century, depending on how fast carbon dioxide emissions grow. Scenarios that assume that people will burn more and more fossil fuel provide the estimates in the top end of the temperature range, while scenarios that assume that greenhouse gas emissions will grow slowly give lower temperature predictions. The *orange line* provides an estimate of global temperatures if greenhouse gases stayed at year 2000 levels (IPCC WG1 AR4 2007)

The amount of future global warming and concomitant problems of Earth's climate change and its effects on living organisms are difficult to impossible to predict. This is one reason the IPCC uses the word projection. Based on our current knowledge, we can project our climate scenarios into the future.

As can be seen from Fig. 2.13 (IPCC AR4 2007), projections are made based on model simulations with assumptions of future greenhouse gas emissions ranging from constant $CO_2$ to high growth (A2) emissions. Based on a range of plausible emission scenarios, average surface temperatures could rise between 2 and 6 °C by the end of the 21st century.

# References and Additional Reading

Archer D (2005) Fate of fossil fuel $CO_2$ in geologic time. J Geophys Res 110:C09505. doi:10.10 29/2004JC002625

Bader DC et al (2008) Climate models: an assessment of strengths and limitations. A report by the US climate change science program and the Subcommittee on Global Change Research. Department of Energy, Office of Biological and Environmental Research, Washington, DC, USA

Bauer JE et al (2013) The changing carbon cycle of the coastal ocean. Nature 504:61–70 (5 Dec 2013). doi 10.1038/nature12857

Bintanja R et al (2013) Important role for ocean warming and increased ice-shelf melt in Antarctic sea-ice expansion. Nat Geosci **6**(5):376–379 (31 March 2013)

Brohan P et al (2006) Uncertainty estimates in regional and global observed temperature changes: a new dataset from 1850. J Geophys Res 111:D12106. doi:10.1029/2005JD006548 (available as PDF)

Brulle RJ (2013) Institutionalizing delay: foundation funding and the creation of US climate change counter-movement organizations. Clim Change. doi:10.1007/s10584-013-1018-7

Charles AJ et al (2011) Constraints on the numerical age of the Paleocene-Eocene boundary. Geochem Geophys Geosyst 12. doi:10.1029/2010GC003426

Cook BI et al (2014) Pan-continental droughts in North America over the last millennium. J Clim 27:383–397. doi:10.1175/JCLI-D-13-00100.1

Cowtan K, Way RG (2014) Coverage bias in the HadCRUT4 temperature series and its impact on recent temperature trends. Q J R Meteorol Soc (12 Feb 2014). doi:10.1002/qj.2297

England MH et al (2014) Recent intensification of wind-driven circulation in the Pacific and the ongoing warming hiatus. Nat Clim Change (9 Feb 2014)

Farmer GT, Cook J (2013) Climate change science: a modern synthesis. Springer, Dordrecht

Feng S et al (2014) Projected climate regime shift under future global warming from multi-model, multi-scenario CMIP5 simulations. Global Planet Change (Jan 2014)

Freidman, TL (2008) Hot, flat, and crowded, Farrar, Straus and Giroux

Fu Q, Johanson CM (2005) Satellite-derived vertical dependence of tropical tropospheric temperature trends. Geophys Res Lett 32(10):L10703. Bibcode:2005 GeoRL.3210703F. doi:10.1029/2004GL022266

Galaasen EV et al (2014) Rapid reductions in North Atlantic Deep Water during the peak of the last interglacial period. Science (20 Feb 2014). doi:10.1126/science.1248667

Haigh JD et al (2010) An influence of solar spectral variations on radiative forcing of climate. Nature 467:696–699 (07 Oct 2010). doi:10.1038/nature09426

Hansen J (2009) Storms of my grandchildren, Bloomsbury. ISBN 978-1-60819-200-7

Hansen J et al (2013) Climate sensitivity, sea level, and atmospheric carbon dioxide. Phil Trans R Soc (A) 371:20120294. doi:10.1098/rsta.2012.0294

IPCC AR4 (2007) Chapter 8: climate models and their evaluation. The IPCC Working Group I, fourth assessment report

IPCC AR5 (2013) Climate change 2013: the physical science basis. Working Group I contribution to the IPCC 5th assessment report

Johnson JS et al (2014) Rapid thinning of Pine Island Glacier in the early Holocene. Science (20 Feb 2014). doi:10.1126/science.1247385

Jones PD et al (2012) Hemispheric and large-scale land surface air temperature variations: an extensive revision and an update to 2010. J Geophys Res. doi:10.1029/2011JD017139

Kosaka Y, Xie S-P (2013) Recent global-warming hiatus tied to equatorial Pacific surface cooling. Nature 501:403–407. doi:10.1038/nature12534

Lavergne C et al (2014) Cessation of deep convection in the open Southern Ocean under anthropogenic climate change. Nat Clim Change. doi:10.1038/nclimate2132

Levitus S et al (2009) Global Ocean heat content 1955–2008 in light of recently revealed instrumentation problems. Geophys Res Lett 36:5

Li X et al (2014) Impacts of the north and tropical Atlantic Ocean on the Antarctic Peninsula and sea ice. Nature 505(7484):538. doi:10.1038/nature12945

Manabe S, Wetherald RT (1967) Thermal equilibrium of the atmosphere with a given distribution of relative humidity. J Atmos Sci 24(3):241–259

Mann ME et al (1998) Global-scale temperature patterns and climate forcing over the past six centuries. Nature 392:779–787

Mann ME et al (1999) Northern Hemisphere temperatures during the past millennium: inferences, uncertainties, and limitations. Geophys Res Lett 26:759–762

Marcott SA et al (2013) A reconstruction of regional and global temperature for the past 11,300 years. Science 339:1198–1201

Morice CP et al (2012) Quantifying uncertainties in global and regional temperature change using an ensemble of observational estimates: the HadCRUT4 dataset. J Geophys Res 117:D08101. doi:10.1029/2011JD017187

NOAA Geophysical Fluid Dynamics Laboratory (2012) NOAA GFDL climate research high-lights image gallery: patterns of greenhouse warming, NOAA GFDL (9 Oct 2012)

Oreskes N, Conway EM (2010) Merchants of doubt: how a handful of scientists obscured the truth on issues from tobacco smoke to global warming. Bloomsbury Press. ISBN 978-1-59691-610-4. merchantsofdoubt.org

Pachauri RK, Reisinger A (ed) (2007) IPCC AR4 SYR Core Writing Team, climate change 2007: synthesis report (SYR), Contribution of Working Groups I, II and III to the fourth assessment report (AR4) of the intergovernmental panel on climate change. IPCC, Geneva, Switzerland. ISBN 92-9169-122-4

Rind DH et al (2014) The impact of different absolute solar irradiance values on current climate model simulations. J Climate 27:1100–1120. doi:10.1175/JCLI-D-13-00136.1

Roe GH, Baker MB (2007) Why is climate sensitivity so unpredictable? Science 318:629–632

Santer B et al (2014) Volcanic contribution to decadal changes in tropospheric temperature. Nat Geosci 7:185–189. doi:10.1038/ngeo2098

Schmidt GA et al (2014) Using paleoclimate comparisons to constrain future projections in CMIP5. Clim Past 10:221–250. doi:10.5194/cp-10-221-2014

Seidel DJ et al (2004) Uncertainty in signals of large-scale climate variations in radiosonde and satellite upper-air temperature datasets. J Clim 17:2225–2240

Shakhova N et al (2013) Ebullition and storm-induced methane release from the East Siberian Arctic Shelf. Nat Geosci 7:64–70

Sokolov AP et al (2009) Probabilistic forecast for 21st century climate based on uncertainties in emissions (without policy) and climate parameters. J Clim 22:5175–5204

Stroeve JC et al (2014) Changes in Arctic melt season and implications for sea ice loss. Geophys Res Lett. Article first published online: 22 Feb 2014. doi:10.1002/2013GL058951

Tarnocai C et al (2009) Soil organic carbon pools in the northern circumpolar permafrost region. Global Biogeochem Cycles 23. GB2023Bibcode:2009GBioC.23.2023T. doi:10.1029/200 8GB003327

Thiede JC et al (2011) Millions of years of Greenland ice sheet history recorded in ocean sediments. Polarforschung 80(3):141–159

Turner J, Overland J (2009) Contrasting climate change in the two polar regions. Polar Res 28(2)

United Nations, Department of Economic and Social Affairs, Population Division (2013) World population prospects: the 2012 revision, key findings and advance tables. Working paper No. ESA/P/WP.227

Wehrli C et al (2013) Correlation of spectral solar irradiance with solar activity as measured by VIRGO. Astron Astrophys 556 (Aug 2013)

Westerhold T et al (2008) New chronology for the late Paleocene thermal maximum and its environmental implications. Palaeogeogr Palaeoclimatol Palaeoecol 257:377–403

Zou C (2006) Recalibration of microwave sounding unit for climate studies using simultaneous nadir overpasses. J Geophys Res 111:D19114. Bibcode: 2006JGRD.11119114Z. doi:10.102 9/2005JD006798

# Appendix A
# Geologic Time

See (Fig. A.1).

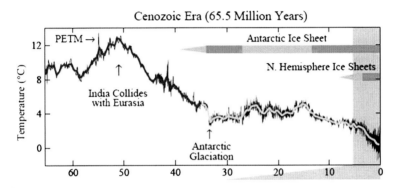

**Fig. A.1** The Cenozoic climatic trend (continuous jagged line from *left* to *right*). Climate change during the last 65 million years as expressed by the oxygen isotope composition of benthic foraminifera. The Paleocene-Eocene Thermal Maximum (PETM) is characterized by a brief but prominent negative excursion, attributed to rapid warming. Note that the excursion is understated in this graph due to the smoothing of data. (Figure prepared by Robert A. Rohde for Global Warming Art) (GFDL)

© The Author(s) 2015
G.T. Farmer, *Modern Climate Change Science*,
SpringerBriefs in Environmental Science, DOI 10.1007/978-3-319-09222-5

# Periods and Epochs of the Cenozoic Era (Fig. A.2)

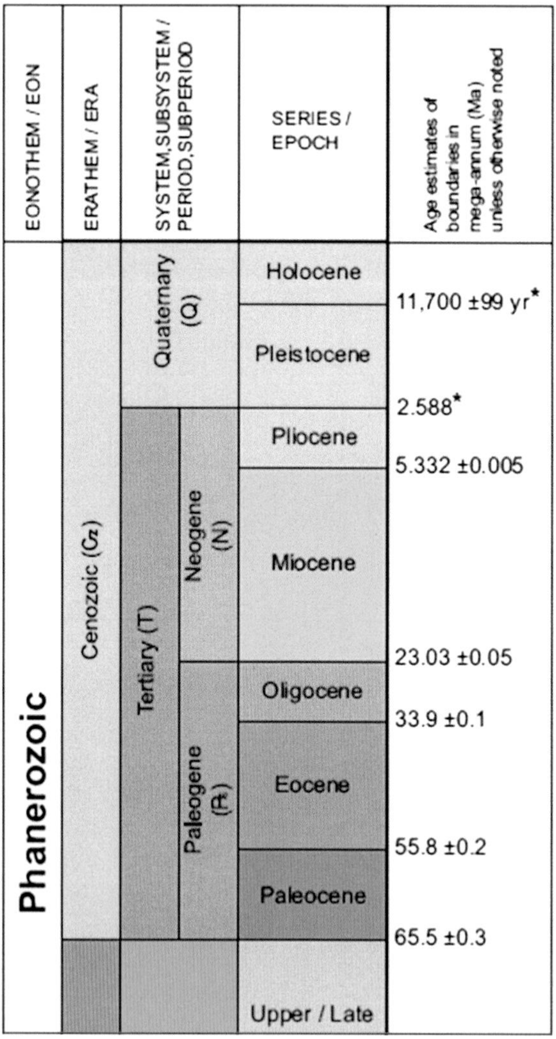

Fig. A.2 Epochs of the Cenozoic Era, Tertiary and Quaternary Periods. (From the USGS, Public Domain)

| EONOTHEM / EON | ERATHEM / ERA | SYSTEM,SUBSYSTEM / PERIOD,SUBPERIOD | | SERIES / EPOCH | Age estimates of boundaries in mega-annum (Ma) unless otherwise noted |
|---|---|---|---|---|---|
| Phanerozoic | Cenozoic (Cz) | Tertiary (T) | Quaternary (Q) | Holocene | |
| | | | | | 11,700 ±99 yr* |
| | | | | Pleistocene | |
| | | | | | 2.588* |
| | | | Neogene (N) | Pliocene | |
| | | | | | 5.332 ±0.005 |
| | | | | Miocene | |
| | | | | | 23.03 ±0.05 |
| | | Paleogene (P) | | Oligocene | |
| | | | | | 33.9 ±0.1 |
| | | | | Eocene | |
| | | | | | 55.8 ±0.2 |
| | | | | Paleocene | |
| | | | | | 65.5 ±0.3 |
| | | | | Upper / Late | |

**The Geologic Time Scale**
**Geologic Periods From Cambrian to Recent (Fig. A.3)**

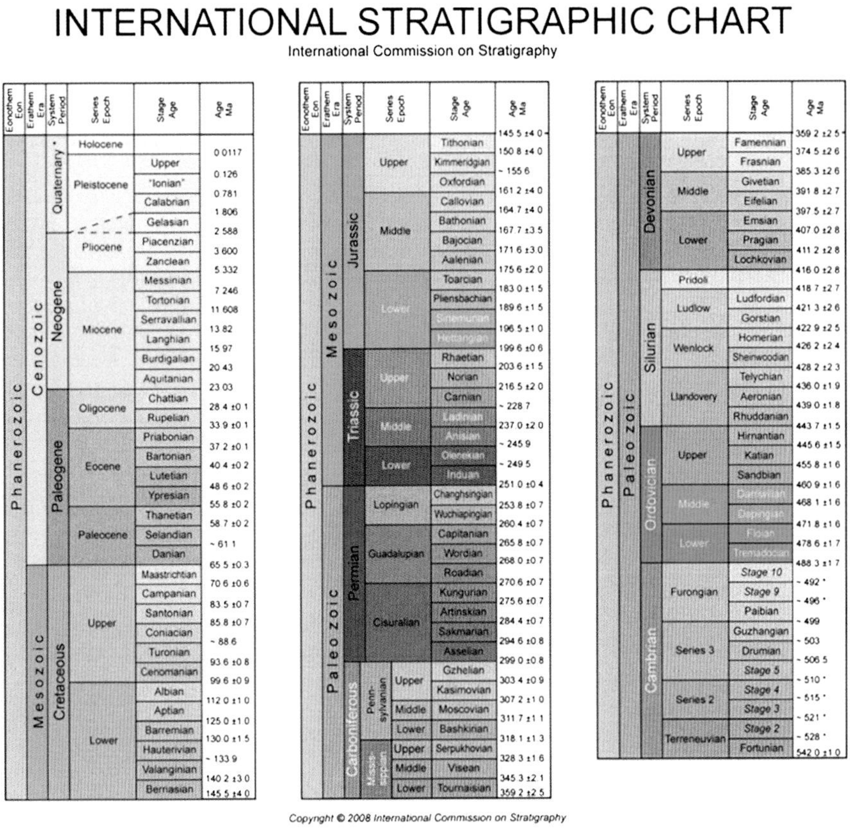

**Fig. A.3** The official geological time scale. (From the International Commission on Stratigraphy)

## Late Pleistocene Glacial-Interglacial Episodes (Fig. A.4)

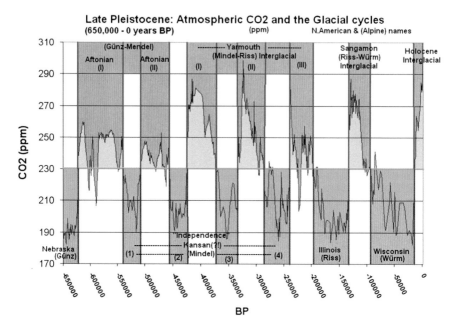

**Fig. A.4** Glacial and interglacial cycles as represented by atmospheric $CO_2$, measured from ice core samples going back 800,000 years. The stage names are part of the North American and the European Alpine subdivisions. The correlation between both subdivisions is tentative. (From Tomruen, Wikimedia Commons)

# Appendix B
# The Metric System

| To convert from | To | Multiply by | More precisely, multiply by |
|---|---|---|---|
| Acres (US survey) | Hectares (ha) | 0.4 | 0.4046873 |
| Feet (ft) | Meters (m) | 0.3 | 0.3048 |
| Fluid ounces (fl oz) | Milliliters (ml) | 30 | 29.57353 |
| Gallons (gal) | Liters (l) | 3.8 | 3.785411784 |
| Inches (in) | Centimeters (cm) | 2.54 | 2.54 |
| Knots | Kilometers per hour (km/h) | | 1.852 |
| Miles (mi) | Kilometers (km) | 1.6 | 1.609344 |
| Miles per gallon (mi/gal) | Liters per 100 km (l/(100 km)) | | Divide 235.215 by mi/gal |
| Miles per hour (mi/h) | Kilometers per hour (km/h) | 1.6 | 1.609344 |
| Nautical miles | Kilometers | | 1.852 |
| Ounces (oz) | Grams (g) | 28 | 28.34952 |
| Pound-force (lbf) | Newtons (N) | | 4.448222 |
| Pounds (lb) | Kilograms (kg) | 0.45 or divide by 2.2 | 0.45359237 |
| Pounds per square inch ($lbf/in^2$) | Kilopascals (kPa) | | 6.894757 |
| Quarts (qt) | Liters (l) | 0.9 | 0.946352946 |
| Square feet ($ft^2$) | Square meters ($m^2$) | 0.1 | 0.09290304 |
| Square miles ($mi^2$) | Square kilometers ($km^2$) | 2.6 | 2.589988 |
| Yards (yd) | Meters (m) | 0.9 | 0.9144 |

© The Author(s) 2015

G.T. Farmer, *Modern Climate Change Science*,
SpringerBriefs in Environmental Science, DOI 10.1007/978-3-319-09222-5

| Quantity measured | Unit | Symbol | Relationship | |
|---|---|---|---|---|
| Length, width, distance, thickness, girth, etc. | Millimeter | mm | 10 mm | =1 cm |
| | Centimeter | cm | 100 cm | =1 m |
| | Meter | m | 1 m = 100 cm | |
| | Kilometer | km | 1 km | =1,000 m |
| | Gram | g | 1,000 g = 1 kg | |
| | Kilogram | kg | 1 kg | =1,000 g |
| | Petagram | Pg | $1,000,000,000,000,000 \text{ g} = 10^{15}$ | |
| | Metric ton | t | 1 t | =1,000 kg |
| Time | Second | s | | |
| Temperature | Degree Celsius degree Fahrenheit | °C °F | | |
| | Square kilometer | km$^2$ | 1 km$^2$ | =100 ha |
| Volume | Milliliter | ml | 1,000 ml | =1 l |
| | Cubic centimeter | cm$^3$ | 1 cm$^3$ | =1 ml |
| | Liter | l | 1,000 l | =1 m$^3$ |
| | Cubic meter | m$^3$ | | |
| Speed, velocity | Meter per second | m/s | | |
| | Kilometer per hour | km/h | 1 km/h | =0.278 m/s |
| Density | Kilogram per cubic meter | kg/m$^3$ | | |
| Force | Newton | N | | |
| Pressure, stress | Kilopascal | kPa | | |
| Power | Watt | W | | |
| | Kilowatt | kW | 1 kW | =1,000 W |
| Energy | Kilojoule | kJ | | |
| | Megajoule | MJ | 1 MJ | =1,000 kJ |
| | Kilowatt hour | kW h | 1 kW h | =3.6 MJ |
| Electric current | Ampere | A | | |